Practical Manual
of Land Development

Practical Manual
of Land Development

Barbara C. Colley, P.E.

Third Edition

McGraw-Hill

New York   San Francisco   Washington, D.C.   Auckland   Bogotá
Caracas   Lisbon   London   Madrid   Mexico City   Milan
Montreal   New Delhi   San Juan   Singapore
Sydney   Tokyo   Toronto

Colley, B. C. (Barbara C.)
    Practical manual of land development / B.C. Colley. — 3rd ed.
        p.   cm.
    Includes bibliographical references and index.
    ISBN 0-07-011967-8
    1. Civil engineering—Handbooks, manuals, etc.   2. Building sites—
Handbooks, manuals, etc.   3. City planning—Handbooks, manuals,
etc.   I. Title.
TA151.C65   1998
624—dc21   98-34765
                                                                       CIP

McGraw-Hill

A Division of The McGraw·Hill Companies

1 2 3 4 5 6 7 8 9 0   DOC/DOC   9 0 3 2 1 0 9 8

ISBN 0-07-011967-8

*The sponsoring editor for this book was Larry Hager, the editing
supervisor was Christina Palaia, and the production supervisor was
Sherri Souffrance. It was set in Century Schoolbook by North Market
Street Graphics.*

Printed and bound by R. R. Donnelley & Sons Company.

McGraw-Hill books are available at special quantity discounts to use as
premiums and sales promotions, or for use in corporate training programs. For more information, please write to the Director of Special
Sales, McGraw-Hill, 11 West 19th Street, New York, NY 10011. Or contact your local bookstore.

This book is printed on recycled, acid-free paper containing a
minimum of 50% recycled de-inked fiber.

To my mother
Alice Mae Duncan
who told me as a child, "You like to talk so much, you should be a writer."

And in memory of my father
Irvin S. Duncan who told me often, "Complications are just misarrangements of simplifications."

Contents

Preface

I recently was given the opportunity to work on a complex project using CADD and learned a great deal from that experience. I have come to believe that the most efficient use of CADD is not obvious. The application of how best to make use of the computer must be carefully considered. I have concluded that preliminary engineering should make use of CADD in a limited way and that the use of the CADD-generated preliminary engineering should have limitations.

The importance of having company policies and standards on organizing computer information is imperative. I have addressed that need in Chapter 11, The Final Plans, Specifications, and Estimates. Taking the time to establish policies and procedures is not simply efficient, it is essential. Those who do not take the time will be unable to compete with those who do.

In 1991, President Bush signed an executive order that all agencies using federal money would be required to begin using the metric system by October 1996. This was the first step to having the metric system come into universal usage in the United States. For this reason this book now uses the metric system as the primary units of measurement. Values in the English system are given parenthetically. Where resources are not now available in metric units, English units are shown.

Use of the Internet is getting a great deal of press. What its value will be to civil engineers has not yet been established, but clearly it will have an impact. It is possible to use the Internet to download governmental documents and construction product details and specification. At this writing, however, the consensus seems to be that it has been most useful in recruiting young engineers by having a web site. As those new recruits mature, they will bring along expanded use of the Internet.

A comment I am hearing from some of my colleagues is that the young engineers are now so engaged in the development of their skills for using CADD that they are not learning the engineering design as quickly as before. This is a problem for the company and for the young engineers that must be addressed. To overcome this problem, one engineering firm now makes a requirement of six months design experience before CADD use is allowed. Another requires reading of the *Practical Manual of Land Development* as a first assignment of engineers coming into the firm.

The purpose of this book is to provide a ready resource for civil engineers putting their education to work in practical applications. For that reason, the content has been limited to those tasks likely to be encountered during day-to-day design tasks by civil design engineers in land development. The term *practical* in the title describes what is perhaps the most significant objective of this book. It has intentionally been kept as concise as is reasonable. The intent is that the user will not have to search through information that, though interesting, is not essential to day-to-day design practice in order to receive the information needed to perform required tasks. I believe that another advantage of this book is that it satisfies the term in the title *manual*—of or by hand; a small book giving instructions. Believing that larger books are cumbersome and more difficult to use, this book has intentionally been kept small.

—*Barbara C. Colley, P.E.*

Acknowledgments

There are a great many people who have helped me in this endeavor who I wish to thank. In particular, I wish to thank Arminta Jensen, Felix Jacob, and Mike Taylor of Rugeri, Jensen, Azar & Associates for their contribution of illustrations of maps and plans using metric calculations. Those illustrations make a significant contribution for the discussion of metrics. Ruth and Going, Inc. are owed a great debt of gratitude for providing me with the opportunity to work on complex site plans using CADD and for their contribution of the Record of Survey Map. The knowledge learned while working with them has been priceless. Michelle Self of CAD Masters and her staff have been most generous with their time and knowledge about Autocad® and Softdesk® products as well as other aspects of working in CADD systems. Without their ready support, the discussions in this book about the use of CADD systems could not have occurred.

Practical Manual
of Land Development

1

Land Development

Making the environment more useful and comfortable for humanity is the purpose of civil engineering. Civil engineering for land development includes the design and construction of transportation corridors; flood control facilities; potable water supply facilities; collection and treatment facilities for solid and waterborne waste products; electrical, gas, and communications facilities; and buildings.

Implementing the development and improvement of land involves political, economic, and aesthetic considerations as well as engineering realities. A project may involve entrepreneurs, financiers, politicians, public agents, architects, landscape architects, geologists, hydrologists, environmentalists, and construction contractors as well as engineers.

The skills and talents of land surveyors, mechanical engineers, and electrical and lighting engineers will be needed in addition to those of several kinds of civil engineers. Civil engineers specializing in traffic, structures, soils, and hydraulics may be needed. Effective communication among them is essential. Lack of clear communication can be the greatest obstacle to timely, satisfactory completion of any project. The intent of this book is to present a clear description of the engineering tasks and to promote a better understanding among the various people involved in land development.

Using This Book

The engineering design of public works and private projects should be done under the supervision of a highly educated, experienced engineer. This book has been written as an overview and guide to the engineering design of a variety of land development projects. The design of each aspect of the project must be made with an understanding and respect for the other aspects.

The information found here is necessarily presented in a broad but shallow way. Readers desiring more depth of understanding should refer to the references at the end of each chapter for further reading on the subject. Work through each of the examples presented in the chapters: the examples contain information about techniques and procedures that are not described in the text. By solving the problems in the examples, the text will become more clear and you will be more likely to retain the information. Reading and solving the problems at the ends of each chapter will show you which information the author considers to be most important.

Nomenclature

Terms used to describe governing agencies, construction materials and techniques, and maps and plans vary in different parts of the country. The usage in this book should make the meaning clear. Great care has been taken to define terms and jargon when first used. However, if the meaning of a word used is not clear, refer to the glossary. The terms *jurisdiction* and *agency* are used frequently and interchangeably throughout this book. They refer to the political body which has power of approval over the aspect of the design being discussed. The jurisdiction may be anything from participants in a town meeting to representatives of the federal government. The terms *pipe, conduit, main, sewer,* and *drain* are also used interchangeably. The term *developer* can refer to a private party, a development company, or a public agency.

Local Customs and Resources

The words chosen to describe materials or procedures in this book may vary from the terms used for the same material or procedures in another part of the United States or another country. It is best to use what is customary locally—unless there is clear evidence that some new terminology, material, or technique is superior. There are always those who resist change, and change initially requires additional time. Local agencies should be consulted for design criteria and specifications. When local agencies have not established criteria, nearby agencies with similar conditions and history or respected contractors working on the area can be helpful. This book is written as a guide only—not as a set of rules.

Coordination

Each aspect of the improvement of any site must be coordinated with every other aspect. One may design the sanitary sewer with no problems, only to discover that its location creates a problem in the design of the storm drain. After both have been redesigned, it may be discovered that the new design creates a problem in a third area. The engineering may go smoothly, only to have the client or a public agent request redesign. The plans must be polished and repolished before they are finished.

No subject or chapter in this book should be used without the others. Each chapter necessarily focuses on one aspect of the improvements, but all aspects are inextricably bound together.

Public Agencies

Every project requires acquisition of permits from agencies charged with protecting the health and welfare of the public. These agencies have established certain criteria and standards. Ordinances have been written and established as law by political processes. Failure to obtain approvals may mean dismantling structures and/or financial penalties. It is right and necessary for public agents to examine plans and to require changes deemed necessary.

The role of the public agent

Public agents have a different perspective than the developer and engineer. They see not only the project but its impact on the immediate neighbors and the community at large. The region must be protected from disturbance of ecosystems. Air, noise, soil, and water pollution must be prevented. The agencies are responsible for verifying that the sewage transportation and treatment facilities are adequate, that the project is not situated where it will be endangered by floodwaters or landslides, and that the existing or planned storm drainage facilities will be adequate to handle increases in storm water runoff that may result from the project. Further, most projects impact traffic flow patterns, requirements for fire and police protection, and possibly school enrollment. These issues should be considered *before* a project is approved, while problems can be avoided or mitigated.

Dealing with public agents

It is important to establish a relationship of mutual cooperation and respect with public agents, whether they be the mayor of a metropolitan area or a file clerk in the county recorder's office. We are dependent on these people for their approvals and assistance.

File clerks may have more valuable information in their heads than all the microfiche and computers in the office. File clerks who have been responsible for maps for many years can be worth their weight in gold. Very old maps and plans may be impossible to find without such people.

The truism that contacts are essential to success is demonstrated daily in this business. Always introduce yourself to agents, and tell them who you represent. Presentation of your card will help them remember you. Be courteous and respectful, and you will be remembered. Write down the names and positions of those you meet. Once acquaintance is established, even if just through telephone contact, information will be forthcoming more easily and quickly.

If an agency plans to deny a permit for your project, ask about its concerns. Suggest solutions that meet those concerns and that satisfy the spirit and intent of the criteria.

Public Hearings

Projects to be built with public funds, such as highways and airports, must be planned with notifications to and input from the general public. Public hearings must be scheduled and notification given to ensure the greatest participation by the public. At these meetings, the lead agency presents the plan, and private citizens and special-interest groups are given the opportunity to present their concerns and suggestions. The public meeting provides a forum for the engineers and other professionals to explain how these concerns are being addressed.

Citizens may point out that a planned freeway will create a barrier between their children and the school their children attend. When this problem is brought to light, the need for a pedestrian overcrossing may be apparent. Other citizens may express concerns about noise and air pollution introduced by a new freeway. Experiences and comments from the public can be a valuable asset.

Private projects are also subject to public hearings if they involve creation of new lots or changing existing zoning. Neighbors within a specified distance of the new project are notified of a planning commission, city council, or county board of supervisors' meeting where they will have the opportunity to express their opinions about the project. If the project will change the character of the neighborhood or adversely affect some of the neighbors, ways are investigated and sometimes mandated to mitigate the potential problems.

The Developer

Each project is different from every other project. What is important to one developer may not be important to another. One developer, such as a state or federal agency, may study every detail and know exactly what materials and construction methods are to be used. Another developer may want to take the most economical approach possible. Know what approach the client wants. The most professionally designed project will not lead to further work from that client if the result is not what the client had in mind. Of course, if what the client wants would not be good engineering practice, the developer must be educated and directed to a more acceptable approach.

Health Issues and Toxic Substances

In recent years, engineers and others have become increasingly concerned about toxic substances and their effects on people's health. It has come to national attention that many of the practices of businesses and industries, as well as of ordinary citizens, have caused damage to the environment that will require years to repair.

Asbestos was once a commonly used construction material. It is now thought to cause lung and other cancers and is no longer used in construction. Asbestos used in ceiling tiles in many schools and public buildings is being removed with very tightly controlled methods to ensure workers' safety.

Underground storage tanks for gasoline and other toxic liquids have been installed without precautions against leakage. Not only are those tanks now being replaced with double-walled tanks, but soils testing and installation of monitoring wells to detect leaks are also being required. Determination of the extent of toxic plumes from previous installations and elaborate systems for monitoring wells are being required.

The previous use and purpose of property must now be known and evaluated before a construction project is undertaken. Liability created by previous owners may come with the property. If a piece of property was previously used for a gas station and there were leaks of gasoline into the soil, the new owner may be required to remove the toxic substances from the soil before any construction can begin. Materials stored on the property may also have adversely affected the soil and/or surrounding properties or waterways.

Methods have been developed to treat some pollutants without extracting the soil (in-situ treatment). New methods are being developed that utilize biochemistry and electrokinetic phenomena in soils. When underground pollution is widespread, this may be the only feasible way of correcting the damage. When extracting the pollutant from the soil is feasible, it can mean excavating the soil and dumping it in a Class 3 dump site. The cost of this approach is high, not only because the cost of removing and replacing the soil is expensive, but also because the fees for using a Class 3 dump site are high. Further, simply finding such a dump site may be difficult, and there may be a problem finding a contractor willing to take the risk of dealing with the toxic materials.

The cost of cleaning a site can be so high as to make the property uneconomical to develop. If property is purchased without determining whether pollution is a factor, an unwary owner can be driven into bankruptcy.

Environmental Issues

The environmental impact of land use and development has become a major factor for project planning. An environmental impact report (EIR) may be required locally, and an Environmental Impact Statement (EIS) is required on most federal projects. Even on very small projects, some examination must be performed in order to qualify for a *negative declaration* stating that the project will not impact the environment.

A draft environmental impact report (DEIR) is prepared that addresses all the issues and ecosystems expected to be affected by a project. The draft report is then circulated through various governmental agencies, and the general public is given an opportunity to comment on and raise concerns that may have been missed by the report writer. Concerns of private citizens or political action committees (PACs) can cause a project to be altered or even terminated as a result of concern for such things as loss of habitat for an endangered species such as the spotted owl or a rare flower that grows only on the project site.

Most often, mitigation can be provided to allow a project to go forward, but this does not always happen. The cost of the study in time and dollars must be factored in when a project is proposed. At the time of this writing, the disappearance of wetlands is an issue of major importance, as wetlands provide a unique environment for flora and fauna. In California, for instance, if wetlands are taken out of use because a highway bridge or some other structure will cover them, those wetlands must be replaced at a rate greater than 1:1. The ratio is based on the quality of the proposed wetlands and the amount of time necessary for the new wetlands to become mature and can be as great as 7:1.

Engineering

There is a popular saying among civil engineers that when alligators are snapping at your ass, you forget that you set out to drain the swamp. Private-sector land development is one of the most time-sensitive of all businesses. The pressure to perform multiple tasks quickly can lead to oversights and errors. On any given day you may work on several different jobs. Interruptions for phone calls to solve minor problems or major crises on other jobs make continuity of thought on your primary task difficult. Keep in mind what you set out to do.

There is considerable risk, particularly to young engineers, of over-engineering simple problems. Construction of a retaining wall may solve a landslide problem, but removing the potential landslide material may cost less and make more sense. If a 4-in plastic pipe will handle roof drainage adequately, don't install a 12-in concrete pipe. Be alert to this risk, and check plans and results using common sense. Keep the work simple but complete. Whenever possible, solve the same problem more than one way and compare the results. It takes only minutes to superimpose two or more CADD layers or to lay one hard copy over another on a light table to compare and look for differences. This simple process can quickly point out discrepancies and is well worth the time.

Working in the metric system

In 1991, President Bush signed an order that, beginning in October 1, 1996, all federally funded projects were to be designed and constructed using the metric system. This is the first step to promote conversion of all engineering and construction in the United States to metric. For that reason, this edition of the *Practical Manual of Land Development* has been written using the metric system. Where resources such as existing maps are used, the dimensions are maintained in the imperial system because that is what the engineer can expect to find at this time. The traditional method of measuring and number usage in the United States is called the *imperial system.* The International System of Units referred to as SI (from Le Système International d'Unités) was adopted by the General Conference on Weights and Measures in 1960. It is the international system that is being universally adopted. The conventions used in this book are taken from three sources:

1. *Standard for Use of the International System of Units (SI): The Modern Metric System,* published as IEEE/ASTM SI 10-1997[1]

2. *Metric Practice Guide for Surveying and Mapping,* by the American Congress on Surveying and Mapping[2]

3. *A-3 Metric Primer,* by CALTRANS[3]

A table of conversion factors follows (see Table 1.1), but it is not sufficient to know the conversion factors. There are a number of questions which arise concerning how conversions are to be handled. For instance, if conversion factors are used exactly, a 12-ft-wide traffic lane would become a lane 3.6576 meters wide. This type of usage is referred to as soft conversion. It makes more sense to make the conversion to a more useful/suitable number. When we convert to a suitable number, it is called a hard conversion. For a 12-ft lane converted to metric, CAL-TRANS uses 3.6 meters. An 11-ft lane is converted to 3.3 meters and a 10-ft lane is converted to 3.0 meters. The designations 3.6, 3.3, and 3.0 meters are equivalent to 12-, 11-, and 10-ft lane widths so that they meet the requirements for safety as well as the more precise number, and they are easier to use. You will want to use hard conversions for most of the work you will be doing in land development.

Some commercial products, such as sizes of lumber, are described in a way so that we understand what is meant even though the description cannot be taken literally. A 2 × 4 is understood to be lumber that is approx-

TABLE 1.1   Metric Conversion Factors

LENGTH	1 mm = 0.03937 inches	1 inch = 25.40 mm
	1 m = 1000 mm	1 mm = 0.001 m
	1 m = 39.37 inches	1 inch = 0.0254 m
	1 m = 3.281 feet	1 foot = 0.3048 m
	1 m = 1.094 yard	1 yard = 0.9144 m
	1 km = 1000 m	1 m = 0.001 km
	1 km = 0.6214 miles	1 mile = 1.609 km
AREA	1 mm^2 = 0.000 010 76 sq. ft.	1 sq. ft. = 92 900 mm^2
	1 m^2 = 1 000 000 mm^2	1 mm^2 = 0.000 001 m^2
	1 m^2 = 10.76 sq. ft.	1 sq. ft. = 0.0929 m^2
	1 m^2 = 1.196 sq. yd.	1 sq. yd. = 0.8361 m^2
	1 hectare (ha) = 10 000 m^2	1 acre = 4047 m^2
	1 ha = 2.471 acres	1 acres = 0.4047 ha
VOLUME	1 m^3 = 35.31 cu. ft.	1 cu. ft. = 0.02832 m^3
	1 L = 0.2642 gal	1 gal = 3.785 L
	1 m^3 = 1.308 cu. yd.	1 cu. yd. = 0.7646 m^3
MASS	1 kg = 2.205 lb.	1 lb. = 0.4536 kg
	1 tonne = 1.102 ton	1 ton = 0.9072 tonne
VELOCITY	1 m/s = 3.281 fps	1 ft/s = 0.3048 m/s
	1 km/h = 0.6214 mph	1 mph = 1.6093 km/h
FLOW	1 m^3/s = 35.31 cfs	1 cfs = 0.02832 m^3/s
	1 m^3/d = 264.17 gal/d	1 gal/d = 0.00378 m^3/d
TEMPERATURE	°C = (°F − 32) × 0.56	°F = (1.80 × °C) + 32°

imately 2 by 4 inches, but in fact that is the rough-sawn size and the finished size is smaller. These kinds of items will retain their descriptions.

Linear measurement is based on the meter (39.37 inches) in the metric system. All distances should be shown in meters, kilometers (1000 meters), or millimeters (0.001 meter). Centimeter (0.01 meter) is not used. Convention is to use the symbols for these dimensions not abbreviations. The symbols are meters (m), kilometers (km), and millimeters (mm). Plurals are not shown. The symbols are always shown in vertical text regardless of surrounding text and in lowercase and are *not* followed by a period. Further the symbol is to be separated from the number by one space such as 36 mm, not 36mm. The choice of whether to use meter, kilometer, or millimeter should be made so that the numerical value will be between 0.1 and 1000. Values smaller than a meter should be shown as millimeters. The exception to this is when the numbers are in a table. In that case, the values should all be in the same metric dimension. Do not mix unit names and unit symbols. Where you use a modifier, place it before the dimension such as square meters or cubic meters.

Areas are based on the metric system. Large areas will be in km^2 or, where we would traditionally use acreage, in hectares (10 000 m^2 = 1 hectare). The symbol ha is used for hectares.

Angular measurements for civil engineering are expressed in degrees, minutes, and seconds or as degrees and decimals of a degree as in the imperial system. Other SI applications use the *radian* (rad) which is the angle subtended by an arc of the circle equal to the radius.

Time is expressed in seconds, minutes, and days as in the imperial system. *Velocity* is expressed as meters per second (1 m/s = 3.281 fps) or kilometers per hour (1 km/hr = 0.6214 mph). *Flow* is expressed as cubic meters per second (1 m³/s = 35.31 cfs). Gallons per hour or day is not used. Instead, use m³/s or L/s where L is the symbol for liters. Here the symbol is capitalized, which is an exception to the rule of using lowercase letters. This is because of the likelihood that the lowercase L (l) is easily confused with a 1 (one). The other exception is where the symbol is taken from a proper name such as Watt (W).

Decimal markers are shown with a period in the United States. A comma is used in some countries. For that reason, commas should not be used to set off large numbers into groups of three. Instead, a space is inserted after each set of three numbers such as 1 000 000 or 0.000 001. When there are just four digits, no space is used as in 4300 or 0.0043 (see Table 1.1).

Numbers

Two words that engineers should have a clear understanding of are *accuracy* and *precision*. Accuracy refers to correctness. An answer is accurate (correct) or it is wrong. Precision is a matter of degree. An accurate measurement of 10 m does not tell you the degree of precision. The number 10.01 m is precise to within 10 mm. The measurement is greater than 10.00 m but not as great as 10.02 m. If a more precise measurement is needed, the measurement must be made to within 1 mm or 0.1 mm and so on. A number can be accurate and not be precise. An accurate value of a slope for drainage purposes of 1 percent may be just as useful as the more precise value of 1.03 percent. But a number that is not accurate, though it may be precise, is worthless.

Your work must always be accurate. The degree of precision you select should be based on what is dictated by the jurisdiction, the client, or common sense. If you are designing a grading plan, ordinarily 20 mm is sufficiently precise. The contractor will not be able to construct it closer. For dimensions and elevations of structures, use 10 mm. The degree of precision that can be accomplished may be plus or minus 5 mm, but do not round off or the precision will be 5 mm plus the amount rounded off.

Surveying property lines and preparing subdivision maps requires more precision. The degree of precision depends on the size of the parcel being surveyed. When measuring angles, 1 degree of difference results in 17.452 m (57.25 ft.) of offset difference for every km of length; 1 minute of difference yields 0.29 m (0.95 ft.) of offset difference for every km; 1 second of difference yields 0.0484 m (0.156 ft.) of offset difference for every km. The degree of precision chosen should be based on the distance measured and the precision required for the finished traverse.

In mathematical calculations, there is no advantage in using one value in a formula that is more precise than another value. The degree of precision of the answer cannot exceed the degree of precision of the least precise value in the equation. For example, where the circumference of a circle ($C = \pi d$) is needed and the diameter is given as 50.25 m there is no point in using more than four significant figures for the value of π. Use 3.14 rather than 3.14159265. A value in a formula may be precise to two places to the right of the decimal even though no numbers are shown. For example, the value of 2 may be exactly 2, and the precision then is whatever is required, such as 2.00. Keep this in mind when selecting numbers to use in making calculations.

When making conversions, determine the required precision as a guide to how many digits to retain in the number to be converted. The converted dimension should be rounded to a minimum number of significant digits to retain required precision. If the value is preceded by "not more than," the number should be rounded down.

When you prepare estimates, the quantities should be rounded off to no less than five meters or one square meter. If the estimate is for a large project, the quantities should be rounded even further. Using more than four or five significant figures when making an estimate which will total millions of dollars is inappropriate and misleading.

When design is complete, quantities can be determined exactly. The exact length of curb and square meters of paving will be known and should be used. However, showing values more precise than even meters is meaningless.

Order of magnitude is a term that is frequently heard in engineering. The term refers to the relative size of a number. Checking that the order of magnitude of an answer is correct is particularly important when using computers because use of computers makes the work more abstract. When you have calculated an answer, check if the order of magnitude seems correct. For instance, if you are using the rational formula $Q = 0.28CIA$ (Eq. 9.1) and the values are $C = 0.9$, $I = 2.2$ mm/hr, and $A = 531$ km^2, the answer should be roughly 30 m^3/s ($Q = 0.3 \times 1 \times 2 \times 500$). An answer close to 3000 km^3 or close to 30 km^3 is obviously wrong—it has the wrong order of magnitude.

One technique that will aid in making correct calculations is always to show the dimensions of the numbers being used. Including the dimensions ensures that all the necessary conversion factors have been included. The dimensions must be calculated as well as the numbers. Meters times meters yields square meters—area. Square meters times meters yields cubic meters—volume. If the dimension of your answer is meters to the fourth power, it is wrong, since meters to the fourth power is physically meaningless.

A clear demonstration of this approach is in use of the rational formula in SI. The rational formula is $Q = CIA$. It was developed in the imperial system. There, the dimension for Q is cubic feet per second (cfs), C is dimensionless, I is given in inches per hour, and A is given in acres. However, in SI, we want Q to be in cubic meters per second, I in millimeters per hour, and A in square kilometers. If you label each value with the correct dimensions, the formula yields the formula in SI with Q in cubic meters per second.

$$Q = CIA$$

$$= C\left(\frac{\text{in}}{\text{hr}} \times \frac{\text{hr}}{3600\ \text{s}} \times \frac{0.0254\ \text{m}}{\text{in}}\right) \times \left(\text{ac} \times \frac{4047\ \text{m}^2}{\text{ac}}\right) = 0.0286\ \text{m}^3/\text{s}$$

Computers

Use of computer-aided design and drafting (CADD) is now commonplace in civil engineering design and land development. Use of computers and state-of-the-art software to calculate complex formulas and manipulate large volumes of numbers has become necessary for designers to remain competitive. Once computer skills have been mastered, complicated computations can be made more easily, more quickly, and with less risk of error.

Another advantage of CADD is that several similar alternatives can be designed with relative ease. There are programs designed to make exhaustive calculations for several alternatives to determine the best alternative relative to some aspect of the design, such as volume of earthwork. Determining what criteria or aspect of the design to examine is the challenging choice that no computer can make. The information the computer can provide allows the engineer to make decisions that maximize resources and meet the client's needs more precisely.

It is particularly important that the engineering designer have a clear understanding of the basics of engineering when using computers to solve engineering problems. Each problem is unique. The logic and reasoning is done by the engineer; the computer simply responds to the engineer's direction to perform mathematical calculations or delineate lines. Engineers should develop expertise with engineering tasks before designing with computers. Inexperienced engineers using computers without a sound engineering background may be left wondering if what they are doing is really correct—or worse, they may think that they know what they are doing when in fact they do not.

Engineering college graduates are entering the field thinking that because they can operate the computer software programs designed to facilitate the engineering design that they are doing engineering. That is not the case. Rather they are little more than computer drafters. When they start using the computer before they have experience with engineering design, it slows their training and will set them back in their career goals. Some companies are requiring six months or a year of experience before the computer can be used so that the engineers will have a better understanding of the design process. Colleges educate their engineering students with the elements of engineering but the overview and day-to-day skills are seldom taught. Even when they are, day-to-day experience is needed for a real understanding. This book provides the day-to-day instruction that will facilitate the learning process. In some companies, it is required reading.

Drawings and illustrations

Drawings and illustrations are important tools for engineers. They put the design concept into a tangible form. And a drawing made to scale is sometimes the fastest way to find the answer to a problem. Scale drawings are included throughout this book. When there is an error in a traverse, plotting the coordinate points and connecting them can make the location of the error apparent. Plotting the points of a profile at a vertical scale 10 times as large as the horizontal scale clearly shows the location of any points that do not fit in a straight line or along a smooth curve.

When using illustrations for members of planning commissions, city councils, and the general public, remember, they will not be familiar with drawings made using exaggerated scales. Such drawings may confuse them and cause negative feelings about what you are trying to illustrate. Use such drawings only if they are absolutely necessary, then precede and follow them with drawings of the same situation drawn at a natural scale, where the horizontal and vertical scales are the same. Even better, illustrate the design or problem in three dimensions. This can be done with a scale model or be drawn in three dimensions. Many software programs are available to make such illustrations.

Existing conditions

It is essential that the project engineer visit the site as early as possible. Each person visiting a site sees different things. It may be necessary to visit the site many times as the design progresses. Never accept the elevations and locations of significant existing structures from previously prepared plans without having a survey crew field-check the information. If the topography is not checked, differences may become apparent during construction, and the cost of redesign or reconstruction at that late stage is not worth the risk.

Planning the Project

Planning the overall schedule for the project before the work begins is essential. The first step is determination of the design-critical paths—those aspects of the design on which other aspects of the work depend.

Doing a perfect job on one branch of a project, only to discover that essential information is not available and cannot be prepared in time to meet schedules, can be disastrous (Fig. 1.1).

Criteria

The criteria established by the various jurisdictional agencies must be followed. They may be documented in several ways, including city and county ordinances and standard plans and specifications. During the approval process, each agency involved can stipulate conditions of approval. The client will have criteria, and your employer may have established criteria in the form of company policies. Other consultants, architects, traffic engineers, soils engineers, and environmentalists impose criteria as well. All these criteria should be kept in mind. A list should be made and checked frequently.

For all projects, there are many criteria that affect the design. Some of the criteria, whether they be schedules, clearances, or some other value, cannot be changed and must be adhered to. These criteria are called *critical points*. Other criteria will not be absolute. All critical points must be identified before the work is begun. In most cases, all the criteria can be met, but it is important to identify the critical points and give them priority during design.

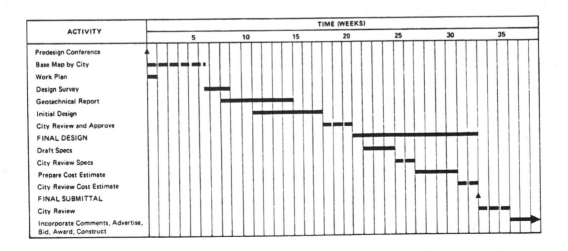

Figure 1.1   Job schedule.

Setback requirements

Planning criteria affects all sites. The architect or planner will have plotted the structures on a drawing of the property taken from the deed or from existing right-of-way or subdivision maps. Whether the new construction is residential, commercial-industrial, or a transportation corridor, there will be minimum distances required between the structures and property lines. Highways and other linear structures have minimum clearances to right-of-way lines and/or other structures. Whether that criteria is met can only be determined with careful calculations.

Planners and architects present their drawings with scaled distances. Engineers and surveyors, however, cannot accept that information. A current *title report* must be acquired from a title insurance company. The title report provides the *grant deed* which is the legal description of the property purchased and lists any property or rights that may have been sold or otherwise granted after the original sale. The title report will provide a legal description of any piece of property that may have been sold from the original purchase, any easements over the property, or any easements attached to the property. Further, the title report will describe other parties who hold interest in the property such as property taxes due, mortgages, and liens.

When the deed is analyzed by the surveyor and a field investigation of the boundary is completed, the distances and bearings may be different from those on the deed. When the true property line is thus established, the distances between structures and the property line may be less than expected and/or is required for setbacks. When this happens, the client and/or architect should be informed immediately, as the design of the structure will have to be altered to comply with the setback requirements.

Easements are rights of limited usage held by one owner over the property of another owner. The most common easements grant a strip of land for vehicular access or for utility lines and their maintenance. Storm drain easements also are not unusual. Engineers must plot the easements that affect the site and read the intent and restrictions imposed by them. Clearance of easements must also be verified by coordinate calculations. Overhangs and paved areas may be described separately, with separate setbacks to be used.

Lot and building sizes and alignment of linear projects are sometimes determined by setback requirements. If several different house plans are designed for a particular subdivision, the engineer may be asked to prepare a *fit list* of which house plans fit which lots. On lots where the fit is questionable, the clearances must be calculated. The setback distances on all projects should be calculated at all critical points.

Timing

The statement "time is money" was never more true than in the land development business. When the economy is booming, land develop-

ment and construction are booming. The faster the projects can be
built, the more money can be made. When economic growth is slow,
many developers are forced out of business. In fast-growing areas, the
political and financial situation can change quickly. The various fees
for building permits or sewer connections can be doubled or tripled
with very little warning. Building moratoriums prohibiting further
development can be imposed at any time. In most areas, wet or cold
weather limits the months suitable for construction. Further, projects
are usually financed and interest costs can be thousands of dollars a
day. These are only some of the reasons why developers often expect
the engineering design to be done quickly. Do not let the time constraints cause you to be careless or to leave out important checking.

Ideally, the topographic and boundary surveys will be complete before
the tentative map is drawn, and the tentative map approvals will be
complete before the parcel or final maps are begun. Sometimes, however, all these processes are started as soon as the contract is signed.
The planners start drawing a tentative map from the deed. The surveyors are sent into the field to collect boundary and topographic information and calculations are started for a final map and drafters begin
work on base maps for the engineering design. Once these tasks are
underway, the engineers and surveyors gather information on the design of existing and proposed utilities in the area. A visit to the site is
essential for the engineer to spot potential trouble spots in the topography. Though an estimate of the existing elevations may be made from
previous projects in the area, the topography of adjacent projects, or
U.S. Geological Survey maps to start the process, careful, thorough field
surveys will have to take place before engineering can begin. Existing
streets and utilities are seldom constructed exactly as designed. *As-
built* plans often are not representative of elevations and locations.

The conditions for approval of the tentative map may require facilities that have not been planned for, and field investigation of the
boundary may reveal that there is less available land than is needed.
The true topography may be quite different than was anticipated.
These differences must be corrected on the plans, adjustments made,
and the deadline still met. Whenever changes are made, the potential
for errors is greatly increased. Utmost care must be exercised to follow
through on every aspect of the design that is affected by changes.

Errors and omissions

Delays during construction to solve engineering problems are very
costly. Equipment and personnel standing idle while the problem is
solved can cost thousands of dollars per hour. Removal and replace-
ment of new structures may be the only solution to a problem that
becomes apparent during construction.

Thoroughness in researching existing and proposed facilities is
essential. Often when design is in progress for one project, design may

also be in progress for an adjacent or nearby property. Remember to ask public agents about other projects that may be planned in the area. Their design and state of completion may affect your project. If the engineer's oversight costs the clients money, you can be sure that they will look elsewhere for an engineer for their next project, or they will look to that engineer for compensation.

Summary

Civil engineering design and land development can be very complex and involve a wide variety of professional people and others. Engineers must understand and work well with the public and public agencies. Clear communication among the participants is essential.

There are many resources available to assist the engineer. The engineer must know how to use those resources wisely. Ultimately, the civil engineering design is a product of those resources and the engineers' experience and sense of logic. Planning the project based on schedules and interdependence of the various aspects of the design is a necessary first step.

Problems

1. What is the greatest obstacle to successful completion of a design project?

2. Name five types of professionals involved in land development.

3. Name three kinds of civil engineers involved in land development.

4. Define *jurisdiction*.

5. When the local jurisdiction has no established criteria for some aspect of the work, what criteria should be used?

6. What is the task of the public agent?

7. What is the purpose of a public hearing?

8. What is the economic impact of a site having been used for toxic or hazardous material storage?

9. What is an EIS?

10. Describe an engineering problem and propose two solutions—one complex and one simple.

11. Convert the following values from the imperial system to SI:
    a. 435.67 lf
    b. 47 miles

      c. 12.22 cy

      d. 22 ac

      e. 27.4 cfs

12. What is the difference between *accuracy* and *precision?*

13. Define *order of magnitude.*

14. Why is it important to use dimensions when solving engineering problems?

15. What is a critical point?

16. Name three sources of criteria for a project.

17. What is a title report? How is it used?

18. Define *easement.*

References

1. IEEE.ASTM SI 10-1997, *Standard for Use of the International System of Units (SI): The Modern Metric System,* Institute of Electrical and Electronic Engineers, Inc., New York, New York, ASTM, West Conshohocken, Pennsylvania.
2. American Congress on Surveying and Mapping, *Metric Practice Guide for Surveying and Mapping,* 1978, 5411 Grosvenor Lane, Ste. 100, Bethesda, MD 20814.
3. California Department of Transportation, *A-3 Metric Primer, The Metric System.*

2

Resources

This chapter describes the types of maps, plans, land surveys, software, and other resources available to civil engineering designers and others involved in land development.

Computers

During the 1980s, advancements in the development of computers and computer software revolutionized the civil engineering profession. Computer systems capable of integrating all aspects of civil engineering design and presenting it in four dimensions (the fourth dimension being time) are now available at a reasonable cost. Even small design firms using personal computers (PCs) can perform most of the same functions and have the work translated into a format that is usable with mainframe systems. The use of computer-aided design and drafting (CADD) systems and other software is included in each chapter where it can be applied.

Installation of navigation satellites (Navstar) encircling the earth has provided a resource only imagined a few years ago. With these satellites and the right receivers, travelers can determine their location anywhere on earth. A receiver is available to the public that can show your location as your vehicle moves through an area shown on a map.

These satellites, used with the right system of receivers, can allow surveyors to locate property and topography on the x, y, and z axes to the degree of precision desired without line of sight and in any kind of weather. This capability can reduce the cost of surveying to a fraction of what it was a few years ago. This method of surveying is referred to as GPS (Global Positioning Systems).

The Internet

The increased use of the Internet in our lives in recent years is phenomenal. What it will mean to civil design engineers and land developers is just beginning to be revealed. An article in *Civil Engineering News* in May 1997 reported that those civil engineering companies that had *web pages,* found that it was most valuable as a resource to recruit young engineers. The survey found that having a web page was not significant to bringing in new clients, so its use for outreach is marginal.

However, public agencies, commercial enterprises providing products for construction, and many other organizations are establishing web pages that can be very useful. From those resources, information such as product specifications, plan specifications, and a wide range of other information can be downloaded. Copying information from one computer system to another is to *download.* These resources are increasing every day. Further, the Internet provides a way that engineers can share information through e-mail and e-mail attachments. At the time of this writing (1998) moving large electronic files from one computer system or Internet location to another computer system may take more time than is efficient. However, that will change before long.

Maps as Resources

Drawings that show the relative locations of various aspects of the physical and legal environment are called *maps.* Maps are usually two-dimensional and show only horizontal relationships. Relief maps are the exception and are three-dimensional, showing the highs and lows of the ground level. Only two-dimensional maps will be discussed in this book. Topographic and aerial maps show physical environments. Topography is included on some of the other maps discussed in this chapter. Parcel maps and subdivision maps serve as legal documents and show unique parcels of land with mathematical exactness and to the exclusion of all other property. Information shown must be such that the location of the property lines is indisputable.

Modern land development would be impossible without maps. Street maps, aerial maps, assessors' parcel maps, zoning maps, flood-zone insurance maps, fault zone maps, enterprise zone maps, U.S. Geological Survey (USGS) maps, U.S. Coast and Geodetic Survey (USC&GS) maps, hazardous materials zone maps, and assessment district maps are some of the kinds of maps used by engineers involved in land development. Some of these maps are now, or will soon be, available on compact disk or on the Internet for designers to install into their computer systems to facilitate their use.

Street maps

When the word *map* is used, most people think of a street or road map. Everyone has used these maps; they are essential when traveling in unfamiliar areas. Street and road maps can be purchased at gas stations or acquired at city and county offices or chamber of commerce offices. Automobile clubs such as the Automobile Association of America (AAA) offer excellent street and road maps.

Aerial maps

Aerial photographs are particularly useful in the planning stages of projects. A single photograph may cover a very large area, or a series of photographs can be used to cover linear projects such as highways and canals. The engineer can see the entire area under consideration and can then better select alternative routes for a highway or a site for a development. The engineer then uses this information to locate the project where it can be built with the least disturbance to the existing conditions—and thus the least cost. These maps are also useful when making public presentations, as they provide information that the general public understands.

Photographs are sometimes reproduced onto Mylar, either alone or with a profile grid so that the project can be designed using the photograph for a base. Photographs can also be digitized for use in computer systems.

U.S. Geological Survey (USGS) maps and U.S. Coast and Geodetic Survey (USC&GS) maps

USGS and USC&GS maps can be very useful in the early stages of research for a project. These maps are topographic maps showing large areas (Fig. 2.1). Because of the scale, only the general topography can be shown. That is, areas may be shown as occupied generally by buildings or orchards or open fields. Other information, such as the existence of a well on the site, may also be shown. In areas of rapid growth, this information may be quickly outdated, so the engineer should determine when the map was prepared. The maps show *contours* at 40-ft intervals, which should not have changed significantly over time. In lieu of more specific information, these maps can be useful.

Zoning maps

The purpose of zoning maps is to show zoning districts (Fig. 2.2). These districts are used to control population densities and the character of growth. The intent is to have compatible land uses adjacent and incompatible uses separate. Generally, the types of zones are industrial, commercial, business and professional, and residential. The zones are further subdivided to define densities and uses. The sizes of lots and required building setback distances for specific zones are described in city ordinances. Zoning can be used for almost any purpose. Historic preservation zones, redevelopment zones, and green belt zones are three.

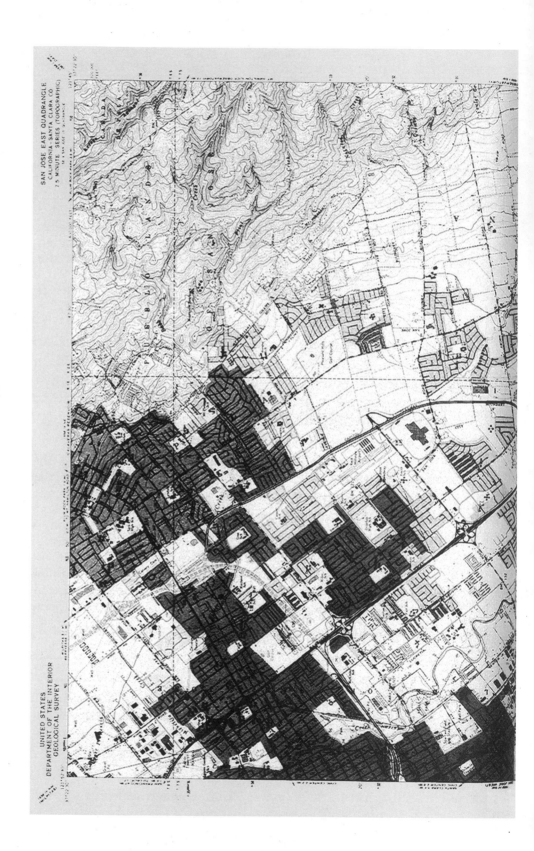

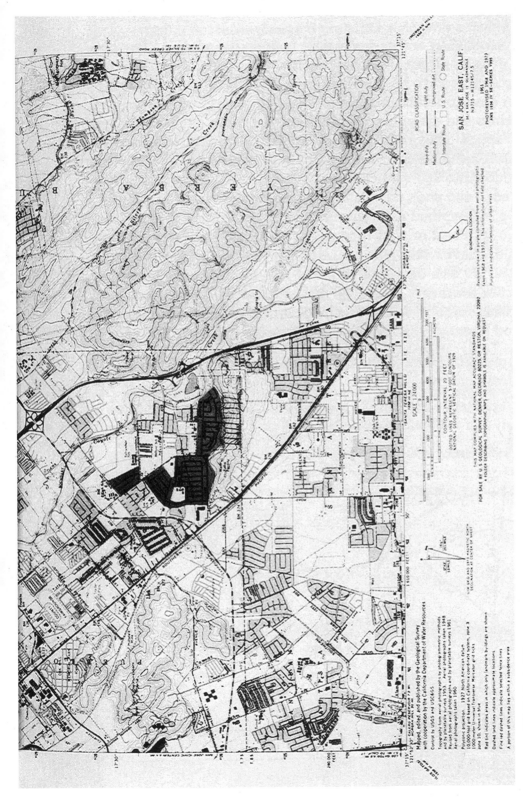

Figure 2.1 U.S. Geological Survey map.

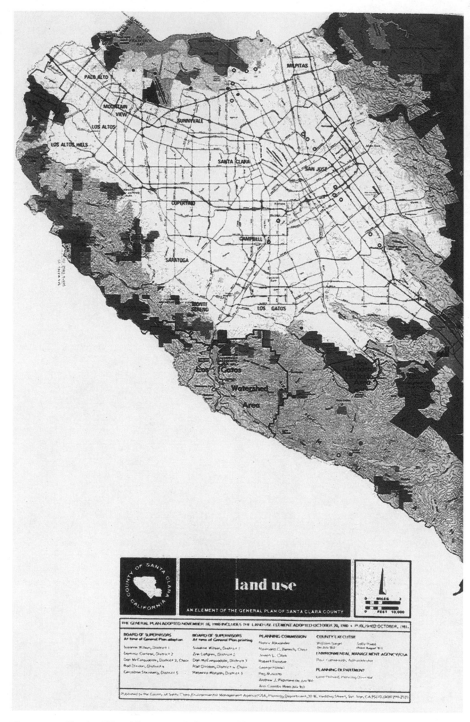

Figure 2.2 Zoning Map (*Courtesy of the city of San Jose, California.*)

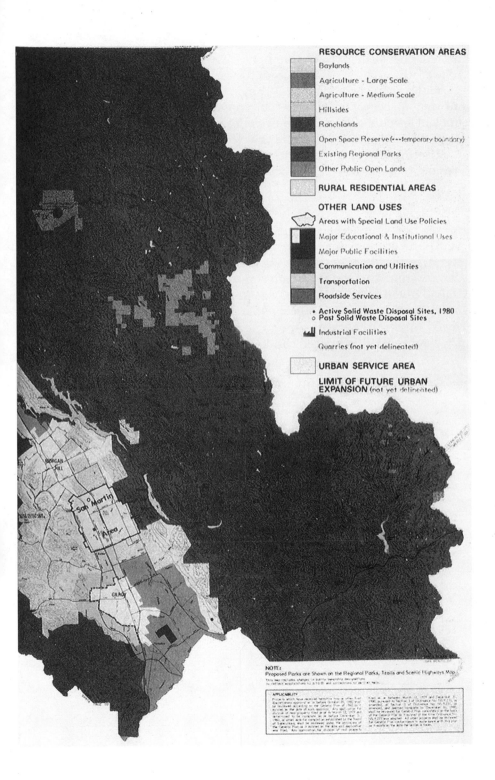

RESOURCE CONSERVATION AREAS

Baylands

Agriculture - Large Scale

Agriculture - Medium Scale

Hillsides

Ranchlands

Open Space Reserve (----temporary boundary)

Existing Regional Parks

Other Public Open Lands

RURAL RESIDENTIAL AREAS

OTHER LAND USES

Areas with Special Land Use Policies

Major Educational & Institutional Uses

Major Public Facilities

Communication and Utilities

Transportation

Roadside Services

• Active Solid Waste Disposal Sites, 1980
○ Past Solid Waste Disposal Sites

Industrial Facilities

Quarries (not yet delineated)

URBAN SERVICE AREA

LIMIT OF FUTURE URBAN EXPANSION (not yet delineated)

NOTE:
Proposed Parks are shown on the Regional Parks, Trails and Scenic Highways Map.

APPLICABILITY

Before work is begun, compliance of the project with zoning requirements must be verified. Deviation from existing zoning can sometimes be allowed through political process. Obtaining such variances, however, usually requires a significant amount of time. Planned developments (not to be confused with development plans) can sometimes circumvent specific ordinance requirements if the developer can demonstrate that there is good cause and that the essence of the zoning is upheld. Mixed-use zoning, wherein commercial, business-professional, and residential uses are placed together, is sometimes allowed.

Assessors' parcel maps

One of the first sources of information for engineers is assessor's maps (Fig. 2.3). The original purpose of assessor's maps is to locate and show property for the purpose of tax assessment. These maps cover all of the land within the county. Traditionally, these maps were filed in books and given page numbers. Assessor's maps are available in the assessor's or tax collector's office, usually located in the county courthouse. By knowing the cross streets near the site, the book where the map of the property is located can be determined from an index map.

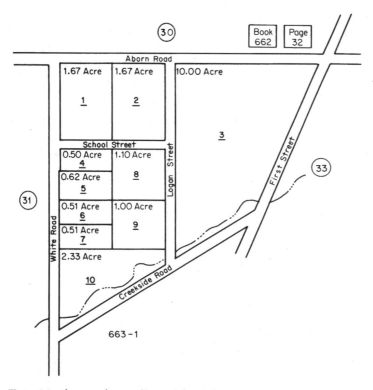

Figure 2.3 Assessor's map (*imperial units*).

Within the assessor's book is another index map to help locate the page (map) that actually shows the property. Reference is given at the edges of the page to show which book and page shows the adjacent properties.

Each page is delineated to show property lines for several properties, and each property is given a lot number. Each property on the tax roles is identified with a number called the *assessor's parcel number* and referred to as the *APN*. The number indicates book-page-lot, such as 662-32-3. This property is shown in Book 662 at Page 32 as Lot 3. The number is then listed in another book, where the name and address of the owner, the zoning, the assessed valuation of the property, and other information is given. In most counties, this information and the assessors' parcel maps have been converted to microfilm or microfiche, or scanned into a computer for easy retrieval and research.

When the maps are copied to microfiche, their sizes change. The change may result in an odd size that cannot be used with conventional measuring instruments. The result is that when the map is copied, the engineer may not be able to scale (measure) the drawing. Drawing scales on the maps facilitate the expansion or contraction of the map to a useful scale.

Another factor that makes the use of assessors' maps for large or linear projects difficult is that adjacent assessors' maps may be of very different scales. In some cases, because of erratic timing and sizes of land developments, one assessor's map may be inside another. The first step in determining ownership for a strip project is to change all the assessors' maps in the affected area to the same scale and then to fit them together like a giant puzzle (Fig. 2.4).

Other information may be shown on assessors' maps. The names and file numbers (tract map numbers) of existing subdivisions and Record of Survey maps in the area may be given. The reference will be shown as a book and page in the Book of Maps filed at the recorder's office. These maps will also probably be available on microfiche. Be alert to the fact that the records may not be current. Find out when the files were last updated, and be certain that you have the most current information. In the future, all this information should be available on a geographic information system (GIS described later) through use of e-mail and the Internet. Engineers will have the most current information at their fingertips, without leaving their offices.

Distances along property lines may be shown on assessors' parcel maps, but these are *not* legal dimensions and should be used only as approximations. True distances can be determined only from the legal descriptions found in deeds filed in the county recorder's office or from title reports of the properties. Street names and widths may be shown, as well as some easements and rights-of-way.

The information on the assessor's map provides a useful sketch for following the legal description. It is of a scale and size that is easily car-

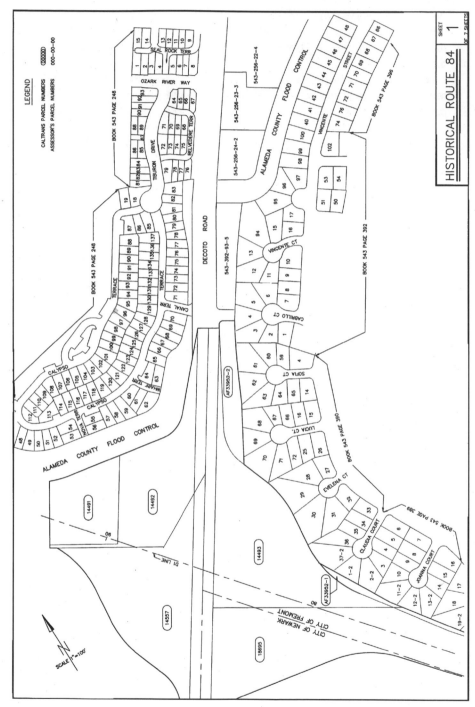

Figure 2.4 Preliminary right-of-way map, composite of assessor's parcel maps.

ried and referred to. It also provides a map that can be given to the surveyors and other consultants working on the project in the early stages of development. With the reference numbers for subdivisions and Records of Surveys adjacent to the site, existing maps and improvement or construction plans can be located. Remember, assessors' parcel maps are graphic representations only. They do not show actual, legal property lines.

Geographic information systems

The information in a *geographic information system (GIS)* is composed of maps and plans of existing and proposed improvements that are selected to be integrated for some stated purpose. The information may consist of geographic or topographic information, including street maps, property and zoning information, existing and planned utilities locations, transportation systems, or any other information deemed useful. This information is integrated into a single source in a computer system. This technology will provide a resource for a vast amount of information that can be made current within hours. Though this resource is still in its infancy, and setting it up will require significant commitment in terms of time and expense, the technology is available and it is only a matter of time before it is readily available to engineers everywhere.

Locating Land

Before construction can begin, a clear and exact location of the property or right-of-way boundary must be established. The deed is the legal description of the property. Its interpretation should be made by a qualified professional (licensed) land surveyor (PLS) from a field survey.

Title reports

Title reports are documents prepared by title insurance companies, which ensure the legal ownership of property. Title reports provide the legal description as taken from deeds and list any claims to the property in the form of easements, mortgages, liens, taxes, and water and mineral rights. Title reports may also show stipulations and limitations on development. Title reports are important to engineers for determining the exact legal description of property. The description of the property can also be found on the deed filed in the county recorder's office, but that information is not sufficient. Subsequent to the original purchase, a portion of the property may have been sold or granted for an easement, or there may be taxes and other monetary interests held by others.

If property is to be developed, its exact property lines and ownership must be known. The title report provides and guarantees the property information. If the land is developed according to where the owners think their property lines are and their neighbor disagrees, there is likely to be a dispute that ends up in the courts. The developer must have a qualified surveyor determine the property lines through a field survey and exacting analysis. The earlier in the project the survey takes place, the less likely there will be a problem with property lines.

Property descriptions

In the simplest cases, a legal description of a property may be, for example, "Lot 1, Tract 5200, filed February 26, 1969, in Book 323 of Maps, at page 65, Santa Clara County Records." In many cases, however, the legal description is more complicated and may continue for pages, describing the property by "metes and bounds" (explained later). If a subdivision map or *monuments* were used in the legal description of the property, the surveyor will have to establish their physical locations. Usually monuments are set at changes in direction of right-of-way lines and at subdivision and lot or parcel corners.

Historically, monuments were significant topographic features described by the deed. At the turn of the century, surveyors would sometimes build their evening campfire at a property or *section corner* (see the following). In the morning, the charcoal from the fire would be buried as the property corner. This was better than installing a wooden stake, because animals would not destroy the monument, and it would be preserved because charcoal does not deteriorate as quickly as wood. Another popular monument used was a mature tree or rock outcropping that could be easily located. Today, ¾-in IPs (iron pipes), tagged with the surveyor's or engineer's license number, are commonly installed along subdivision boundaries. Cities often require that standard city monuments be installed at property corners and along centerlines of roads in new subdivisions.

The following are some of the kinds of elements used in land descriptions.

1. *Subdivision maps.*   Tract 5700, Map of W. E. Woodhams Tract, filed on March 4, 1965, in Book 376 of Maps, at page 31, Shasta County Records

2. *Streets.*   The northwesterly line of Alum Rock Avenue 30.5 m (100 feet) wide

3. *Other deeds.*   The deed from Henry W Smith, et ux,* to George Davis, et ux, dated December 10, 1954, and recorded December 15,

---

* *Et ux* is a Latin abbreviation for *and wife; et vir* means *and husband;* and *et al.* means *and others.*

1954, in Book 303 of Official Records, page 288, Sonoma County Records

4. *Monuments.* A ¾-in IP at the southerly corner of Tract 5200

Metes and bounds. Before a grid system was established to describe land, early settlers described their property by metes and bounds. *Metes* is measure. *Bounds* can be anything, such as a ridge line or creek, that limits the parcel of land to be described. Early settlers described their property as extending from one landmark to another. In some cases the property was then measured by two men on horseback carrying a rope or leather thong between them. The rope was set at 100 varas—0.84 m (33 in). These original property boundaries are still valid, but the description has since been retraced using modern surveying instruments. The boundary is now described using bearings and distances and is called a *metes-and-bounds* description.

When ownership of an irregular plot of land is transferred today, a Record of Survey or Boundary map may be prepared and referred to, but metes-and-bounds descriptions are not uncommon.

Bearings are directions as measured east or west of north or south. They must be exact, and so are measured in degrees, minutes, seconds, and decimals of a second and are determined in the following way. A circle is divided into 360 degrees (360°). Each degree is divided into 60 minutes (60′), and each minute into 60 seconds (60″). The course is described as a certain number of degrees, minutes, and seconds (from 0° to 90°) east or west of north, or a certain number of degrees, minutes, and seconds east or west of south. A line described as N 50°E can also be described as S 50°W. The direction of tracing the course determines whether the bearing is northeast or southwest (Fig. 2.5).

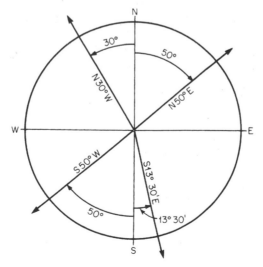

Figure 2.5 Determining bearings.

In a metes-and-bounds description, the bearings and distances serve to traverse property. The following is an example of the metes-and-bounds description of the property in Fig. 2.6. (Distances are shown parenthetically in imperial units in Fig. 2.6 and in this description to facilitate understanding by the reader during the transition to metric units but should *not* be shown on maps and plans or included in descriptions.)

All that certain real property situate in the City of Redding, County of Shasta, State of California, described as follows:

Beginning at the point of intersection of the Northerly line of Main Street, 15.2 m (50 feet) wide, and the Easterly line of First Street, 12.2 m (40 feet wide); thence along said Easterly line of First Street N5°22′36″E, 61.4 m (201.50 feet) to the Southerly boundary of Tract 5700, filed March 4, 1965, in Book 376 of Maps, at page 31, Shasta County Records; thence along the said Southerly boundary of said Tract 5700, N89°52′22″E, 75.3 m (247.13 feet) to the Westerly line of Second Street, 18.3 m (60 feet) wide; thence along Second Street S5°22′36″W, 52.1 m (170.98 feet); thence along a curve to the right with a radius of 9.1 m (30.00 feet), through a central angle of 83°40′32″ an arc length of 13.4 m (43.81 feet) to a point on the Northerly line of Main Street; thence S89°03′08″W, 67.3 m (220.66 feet) to the Point of Beginning.

Containing 4532 m² (1.12 acres) more or less.

Accurate interpretation or writing of metes-and-bounds descriptions is an exacting skill that is beyond the scope of this book. For more information on this subject, see *Land Survey Descriptions* by William C. Wattles.

Townships and sections.   The United States was surveyed by U.S. government surveyors and a grid system was established in the late 1700s

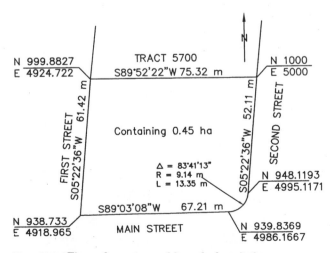

Figure 2.6   Figure for metes-and-bounds description.

and early 1800s as a result of political persuasion by Thomas Jefferson. The land was surveyed and marked into sections. As a general rule, monuments were set at section corners 1.609 km (1 mi) apart and quarter corners 0.805 m (½ mile) apart. Even though the work was sometimes done under very difficult conditions and with instruments that were crude by today's standards, the original monuments, when they can be found, hold the most credibility for determining the location of sections. The original surveys are recorded on government plat maps, and survey (call) notes are available from Bureau of Land Management (BLM) offices. Establishment of townships and sections is a complicated procedure that is beyond the scope of this book. An idealization of the procedure is provided here.

The establishment of townships is based on the latitudes and longitudes of the earth. Longitudes are imaginary lines running north and south through the poles. Each longitude is identified as from 0 to 180° east or west of Greenwich, England. Latitudes are lines extending around the earth, parallel to the equator, and divided into degrees, from 0° at the equator to 90° at the poles.

Prominent geographical points are identified by latitude and longitude, and then referred to for further refinement. Mt. Diablo in Northern California is at latitude 37°51′30″North (the base line) and longitude 121°54′48″West (the meridian line). These are the reference lines for the references in descriptions using Mt. Diablo Base and Meridian, abbreviated MDM. There are two other base and meridian points in California: San Bernardino Base and Meridian and Humbolt Base and Meridian. There are 37 base points throughout the United States.

From these base and meridian lines, townships are established. Townships are rectangular blocks of land 9656 m (6 mi) square. They are described by their distance from the base and meridian. The land contained in the first 9656 m (6 mi) north of the base line is said to be in Township 1 North. The land within the first 9656 m (6 mi) south is said to be in Township 1 South. The land contained in the second 9656 m (6 mi) north is Township 2 North, and so on. The east and west limits of the township are measured in 9656 m (6 mi) increments east and west and are referred to as ranges. The township and range terms are abbreviated. T5N, R2E, MDM (Township 5 North, Range 2 East, Mt. Diablo Base and Meridian) is illustrated in Fig. 2.7.

The township is further divided into *sections* of land approximately 1 mi square, containing approximately 259 hectares (640 acres). The sections are numbered from 1 to 36, starting at the northeasterly corner of the township and continuing back and forth across the township in a zigzag manner (Fig. 2.8). The section is further divided into halves, and quarters, and quarter quarter sections, and so on. The quarters are established by bisecting the boundaries of the section. The points of bisection are called quarter corners. Lines are then drawn between

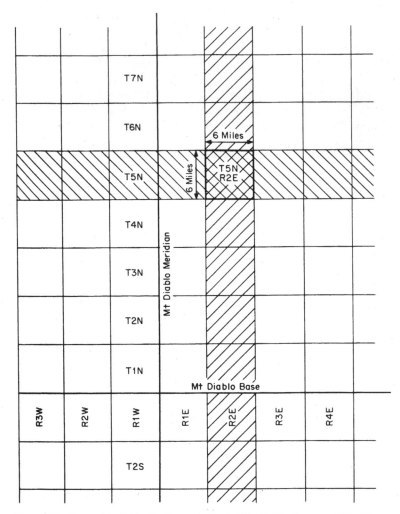

Figure 2.7 Township 5 North, Range 2 East, Mt. Diablo Base and Meridian.

quarter corners to establish quarter sections. The property described as "the North half of the Northeast quarter of the Southwest quarter of Section 26 T5N, R2E, MDM" is shown in Fig. 2.9. To interpret descriptions in this form, start reading at the end of the sentence and trace the process backward.

Example 2.1 Determine the approximate distance between MDM and the west quarter corner of Section 7 T5N, R2E. Use Figs. 2.7, 2.8, 2.9.

solution T5N occupies the land between 38.62 and 48.28 km (24 and 30 mi) north of MDM. Section 7 lies between 0.64 and 0.80 km (4 and 5 mi) north of the south boundary of Township 5N. The west quarter corner is 0.08 km (0.5 mi) north of the section corner.

36	31	32	33	34	35	36	31
1	6	5	4	3	2	1	6
12	7	8	9	10	11	12	7
13	18	17	16	15	14	13	18
24	19	20	21	22	23	24	19
25	30	29	28	27	26	25	30
36	31	32	33	34	35	36	31
1	6	5	4	3	2	1	6

one mile

six miles

Figure 2.8   Township divided into sections.

$$38.62 \text{ km} + 48.28 \text{ km} + 0.08 \text{ km} = 45.87 \text{ km north of the base}$$

$$(24 \text{ mi} + 4 \text{ mi} + 0.5 = 28.5 \text{ mi})$$

R2E lies between 9.66 and 19.31 km (6 and 12 mi) east of the meridian. Section 7 lies between 0 and 1.61 km (1 mi) east of the range line; the west quarter corner is on the west line (0 mi east of the range line).

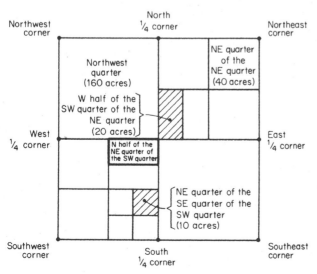

Figure 2.9   Section 26, T5N, R2E, MDM.

$$19.31 \text{ km} + 0 \text{ km} + 0 \text{ km} = 19.31 \text{ km east of the meridian}$$

$$(6 \text{ mi} + 0 \text{ mi} + 0 \text{ mi} = 6 \text{ mi})$$

The west quarter corner of Section 7 is 45.87 km (28.5 mi) north and 9.66 km (6 mi) east of MDM.

The exact dimensions of the distances along the sides of a section and portions of a section will vary from expected dimensions because of inaccuracies in the original surveys. Also, longitudes (meridian lines) converge from the equator to the poles. When adjustment is made for the convergence, the sections along the north and west of the township were made to accommodate the variance.

Coordinate systems

One of the most useful tools at the engineer's disposal is coordinate systems. Use of coordinate systems is great and still increasing with the development of modern surveying equipment and computer-aided design and drafting systems (CADD). A basic understanding of their use is imperative, even though step-by-step procedures can be followed with computers to determine information without understanding the basics. Without a basic understanding, coordinates are meaningless numbers.

Coordinates are numbers that represent distances north and east of a reference point. The reference point can be real, such as Mt. Diablo Base and Meridian, or it can be fictional. Coordinate systems employ trigonometric relationships between points to determine distances and bearings exactly. The use of a coordinate system also facilitates location and plotting of property corners and other survey points with accuracy and precision.

The reference point is at the point of the 90° intersection of a north-south axis with an east-west axis. The north-south axis can be true north, magnetic north, or an assumed north. The point can be given coordinates based on another coordinate system to which ties are made, or it can be given assumed coordinates that are convenient to the task. In Fig. 2.10, point A is the reference point and has been assigned assumed coordinates of 1000.000N and 2000.000E. When traversing a course which is 152.4 m (500.00 ft) long on a bearing of N 50°E from point *A,* the course will end at point *B.* The difference in northerly coordinates is called the *latitude.* The difference in easterly coordinates is called the *departure.* A right triangle is formed by the course, N 50°E, 152.40 m (500.00 ft), the latitude measured along the north-south axis, and the departure as measured from point *C* to point *B.*

From trigonometry we know that the cosine of angle CAB (50°) is equal to the length of $\overline{AC}$ divided by the length of course $\overline{AB}$.

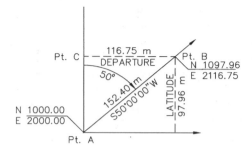

Figure 2.10 A point established using coordinates.

$$\cos 50° = \frac{\overline{AC}}{\overline{AB}}$$

This formula can be manipulated to yield the unknown latitude ($\overline{AC}$).

$\overline{AC} = \overline{AB} \times \cos 50°$ $\overline{AC} = \overline{AB} \times \cos 50°$

$= 152.40 \text{ m} \times 0.642788$ $= 500 \text{ ft} \times 0.642788$

$= 97.96 \text{ m}$ $= 321.39 \text{ ft}$

The northing coordinate for point B is

metric   1000.000N + 97.96 m = 1097.96N

imperial   (1000.000N + 321.394 ft = 1321.394N)

Trigonometry also gives us the relationship that the sine of angle CAB (50°) is equal to the length of $\overline{CB}$ divided by the length of course $\overline{AB}$.

$$\sin 50° = \frac{\overline{CB}}{\overline{AB}}$$

Again manipulating the formula, the departure $\overline{CB}$ can be determined.

$\overline{CB} = \overline{AB} \times \sin 50°$ $\overline{CB} = \overline{AB} \times \sin 50°$

$= 152.40 \text{ m} \times 0.766044$ $= 500 \text{ ft} \times 0.766044$

$= 116.75 \text{ m}$ $= 383.02 \text{ ft}$

The easterly coordinate for point B is

2000.000E + 116.75 m = 2116.75E

(2000.000E + 383.022 ft = 2383.022E)

If the coordinates of points A and B are known but the distances and bearing between them are not, the course between them can be determined. This procedure is called *inversing*. The northerly coordinate of point B (1097.96) minus the northerly coordinate of point A (1000.00) gives the latitude 97.96 m (321.39 ft). The easterly coordinate for point B 2116.75 (2383.02) minus the easterly coordinate of point A (2000.00) gives the departure 116.75 m (383.02 ft.) From trigonometry we know that the tangent of the angle CAB is

$$\tan \angle CAB = \frac{\overline{CB}}{\overline{CA}} \qquad\qquad \tan \angle CAB = \frac{\overline{CB}}{\overline{CA}}$$

$$= \frac{116.75 \text{ m}}{97.96 \text{ m}} = 1.1918 \qquad\qquad = \frac{383.02 \text{ ft}}{321.39 \text{ ft}} = 1.1918$$

The angle whose tangent is 1.1918 is 50°. To find the length of the line, we again use trigonometry:

$$\sin \angle CAB = \frac{\overline{CB}}{\overline{AB}} \qquad\qquad \sin \angle CAB = \frac{\overline{CB}}{\overline{AB}}$$

$$\overline{AB} = \frac{\overline{CB}}{\sin \angle CAB} \qquad\qquad \overline{AB} = \frac{\overline{CB}}{\sin \angle CAB}$$

$$= \frac{116.75}{\sin 50°} = 152.4 \text{ m} \qquad\qquad = \frac{383.02 \text{ ft}}{\sin 50°} = 500.00 \text{ ft}$$

A series of courses is called a *traverse*. The traverse can be continued around a piece of land and back to the point of beginning to delineate property lines such as those shown in Fig. 2.6. An illustration of how to determine coordinates for the property in Fig. 2.6 is given in Fig. 2.11. The traverse starts at assumed coordinates at the northeast corner of the property. The assumed coordinates chosen for that point are large enough that the property will have all positive (not negative) coordinates and small enough that unnecessary numbers will not have to be carried. The number for the easting coordinate was chosen with the same considerations and so that the easting coordinates would not be confused with the northing coordinates. When a traverse contains one or two unknowns but the beginning and ending coordinates are known, the unknown information can be calculated. The unknowns can be as follows:

The bearing and distance of one course

The bearings of two courses

The distances of two courses

The bearing of one course and the distance of another course

TRAVERSE OF FIGURE 2.6							
Point	Distance	Bearing	Cosine	Sine	Northing	Easting	Point
1					1000.00	5000.00	1
	52.11	S 5 22' 36" W	0.995600	0.093703	-51.88	-4.88	
2					948.12	4995.12	2
	9.14	N 84 37'24" W	0.093703	0.995600	0.86	-9.10	
3					948.98	4986.02	3
	9.14	S 0 56' 52" E	0.999863	0.016541	-9.14	0.15	
4					939.84	4986.17	4
	67.21	S 89 03' 08" W	0.016541	0.999863	-1.11	-67.20	
5					938.73	4918.97	5
	61.42	N 5 22' 36" E	0.995600	0.093703	61.15	5.76	
6					999.88	4924.72	6
	75.32	N 89 52' 22" E	0.002220	0.999998	0.17	75.32	
1					1000.04	5000.04	1
		Error of Closure			0.04	0.04	

Figure 2.11   Calculation of coordinates for boundary of property in Fig. 2.6.

Procedures for solving unknowns in two courses using trigonometry are long and complicated, and there are many opportunities for errors. Fortunately, computers and handheld calculators do this task easily and quickly when simple, step-by-step instructions are followed.

You may find that occasionally you will have a problem that the computer cannot solve because of the location of the unknown information in relation to the known information. When this happens, consider that you may be able to solve for unknowns using the geometry and trigonometric formulas learned in high school. It is seductive thinking that the computer can always solve problems most easily. That is not always the case. You may also be able to solve for the unknowns by using some creative thinking about relocating or realigning what is known. One technique that can be helpful is to rotate the traverse so that one course lies due north. Then solving for courses gives you the angles between courses. Knowing angles between the courses, may give you enough other information to solve for unknowns using the computer.

Existing and proposed maps and plans

There are existing maps and plans useful for developing proposed project sites. A description of the kinds of maps and plans that can be used as resources is described at length in Chap. 4.

In the private sector, it is not unusual for work on a desirable piece of land to have been started by one developer and then suspended because of a change in the economy or because of a lack of sufficient planning by the developer. Even though the new owner may have a different concept, there will be valuable information contained on the earlier maps and plans. A thorough search must be made to collect all available resources. When these resources are located, you may find that much of the information you need is already available. After comparing a recent topographic map with observed conditions on a site visit, you may decide that it can be used to begin design as long as you perform a field survey to check critical points. Public works officials are well aware of the effect of time and inflation on the cost of land, so it is not unusual for them to have set aside land through zoning or purchases for transportation systems or other public purposes. New public works projects can be in the planning stages for years before construction actually begins. If a planned public works project may affect your project, you must do a thorough investigation of its impact.

Existing maps and plans may contain a significant amount of information that you will want to use. When this is the case, it may be most cost-effective to have the existing maps scanned electronically and provided to you on a computer disk. This can save hours of drafting time and allow work to begin sooner than otherwise.

Surveys

Engineering for land development depends on professional land surveying. Even though there may appear to be sufficient information as to the topography and property lines from existing maps and plans, the engineer must verify the conditions with a field survey and analysis of the field information. Surface and subsurface topography and property-line monuments are changed through both conscious effort and accidents. Engineers take responsibility for the maps and plans they sign. They must be certain that the information is correct.

Control surveys

Control surveys establish and tie together, with the use of trigonometry, various points on the ground that provide reference points for further surveying. They require a high degree of precision and are used throughout the design and construction of a project.

Boundary surveys

Determination of property boundaries requires sophisticated procedures and specialized expertise. Because of the importance of land ownership both monetarily and emotionally, this is one task that only experienced, licensed land surveyors should perform.

It is not unusual to find a number of similar monuments at the approximate corner of a property. The surveyor must analyze the total

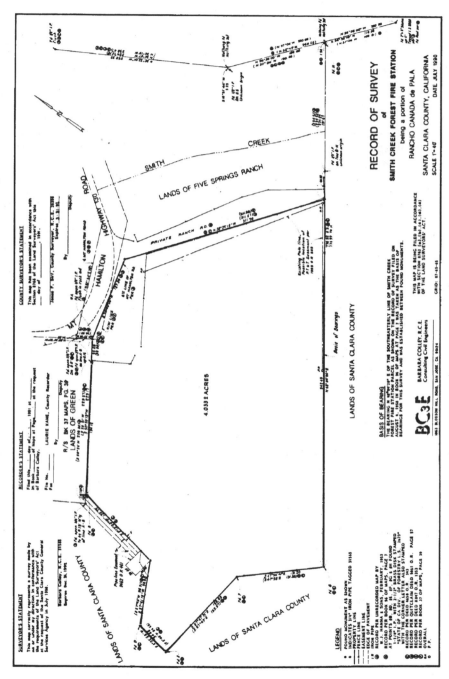

Figure 2.12 Record of Survey map (imperial units)

property boundary and other factors before determining by a prepon-derance of evidence which monument, if any, is correct. When boundary surveys are not performed correctly, overlaps in and gaps between properties may be created, and the chances of a civil suit increase.

Photogrammetry

Photogrammetry is a surveying process that utilizes aerial photographs. By locating survey control points on the ground using conventional sur-veying methods and marking them large enough so that they will be clear on aerial photographs, aerial maps can be produced that can be measured precisely. These photographs can be prepared to whatever scale is most useful to the engineer and to the degree of precision needed. Using complicated procedures and equipment, the photographs can be viewed in three dimensions. Skilled photogrammetrists can then produce maps with contour lines and other topography.

Topography

Each project requires specific topographic information. Topography produced from aerial surveys may not provide information that is specific enough. Aerial surveys also do not provide subsurface informa-tion, such as flow-line elevations on storm and sanitary sewers. Under-ground elevations of water and gas lines can be estimated by locating the elevations of their valves, but more specific information may be necessary. When this is the case, the pipeline must be "potholed," so that the top of the pipe is exposed for the surveyors to measure.

It is particularly important for the engineer to investigate the site. What may show up on an aerial map simply as a building does not explain the importance or cost of the building. Whether the building is a rundown shack or houses multimillion dollar equipment cannot be determined from an aerial survey alone. On the surface, areas where the improvements connect to existing conditions is particularly impor-tant. When connecting to an existing roadway, the minimum informa-tion that must be gathered is elevations up the roadway, and down must be surveyed at the centerline, at the gutter line or flow line, and at the top of curb for a distance of 15 to 30 m (50 to 100 ft) in order to design smooth connections. Further, the next upstream and down-stream catch basins or other drainage facilities must be surveyed in order to determine the areas of drainage basins.

Global Positioning Systems (GPS)

By 1993, 24 satellites were in place at an altitude of 17 700 km (11,000 mi) to broadcast three-dimensional positioning data 24 hours a day to anywhere in the world. This is called Global Positioning Systems (GPS).

The technology has been used since 1979 for navigation and became available to civilians in 1983. It was used in 1984 as survey support for construction of the Stanford Linear Accelerator Extension and was planned for use to support construction of the Superconducting Super Collider in Texas. Measurements can be made to within 1 mm.

GPS surveying transcends many disciplines and requires understanding of orbital mechanics, time, relativity, gravity, mathematics, physics, earth dynamics, and statistics, as well as the limitations caused by the effects of the troposphere and the ionosphere on GPS signals. However, the instruments themselves have been made easy to use and, with proper supervision, surveyors can be operating the antenna and data collectors in a short time.

The use of GPS is clearly indicated for providing control surveys for any large project. Surveying a series of lines to a benchmark while hiking over hills and down canyons, around buildings or other obstacles to the line of sight, or around a bay, can be very time consuming and expensive. With the GPS, the survey can be performed any time, day or night, in any weather, and without line of sight and with much less opportunity for error.

As the technology improves and the cost of instruments drops, use of the GPS will increase. There are now available instruments so that a single surveyor can set up a microstrip antenna attached to a previously established GPS point and walk about a site collecting data with a handheld data collector. The data can later be downloaded for printing of surface and subsurface topography and analyzed for preparation of control and boundary maps.

Working with Computers

The computer revolution of the 1980s has affected the civil engineering profession enormously. The use of computers and state-of-the-art software to perform complex operations and manipulate huge amounts of information has become necessary for designers to remain competitive. Use of computer-aided design and drafting (CADD) is now commonplace in civil engineering design and land development.

Many public agencies currently require the work product of plans and specifications to be submitted on electronic medium using specified software. Software that allows translation from one computer language to another is improving quickly. In most cases, developing the design for a project on one software product, and then presenting it in a format that can be used by another software product can be accomplished. However, when sharing electronic information, whether the platform is DOS, Windows, or some other maker and which release of the software is being used can create problems. Therefore, it is important for the provider and receiver of the electronic information to coordinate that information as well as other information.

A brief description of available computer resources is presented in this section. More specific information on particular software is provided where we discuss the area in which it will be used. Keep in mind that the computer industry is evolving so rapidly that in a very short time additional and better software may be available. If software is of particular interest to you, first check the references section of the chapter covering the subject of interest. If you don't find what you need there, refer to current magazines such as *Civil Engineering, Engineering News Record,* or *Civil Engineering News.*

A word of caution is appropriate here. There is hardly a person who uses computers that has not encountered a computer *virus* at one time or another. When electronic information is being shared, even from very reliable sources, it must be checked for viruses before being installed. Even after taking that precaution, most engineers *backup* or copy computer files to floppy disks, tapes, or compact disks frequently. A virus or other breakdown of the system can incur huge costs to recreate information. What can be even more important is the time lost. Developers are not sympathetic to excuses that you can't meet their schedule because of computer problems. The computer information can be backed up to tape through manual or automatic programs. Two copies should be made and one copy stored at a different location such as in a safety deposit box.

Computer-aided design and drafting (CADD)

The use of computer-aided design and drafting (CADD) software is now being required by many public agencies. Plans prepared on electronic media facilitate storage. New projects prepared this way can also be integrated directly with geographic information systems (GIS).

Information can be taken directly from conventional, aerial, or GPS surveys, digitized, and made available to the design engineer and surveyor in an electronic format. Surveyors can analyze boundary information, apply complicated adjustments with available software, and determine boundaries faster with the use of computers.

When design engineers have set up the design sheet, established the orientation of north, and labeled a starting point with the correct coordinates, the coordinates of other points can be calculated automatically. As a result, lines and arcs can be plotted with ease, and unknowns can be solved with very little effort by the engineer. Engineers can then experiment with different designs and solutions. Further, by plotting the courses, most errors become apparent. CADD systems take away most of the tedious drudgery of complicated calculations and allow the designer much more flexibility in design.

All engineers should acquaint themselves with CADD systems, even if they will not be using them in their day-to-day activities. Using the right software, the engineer can perform the following basic operations with the touch of a button:

Create the following:

a line of specified bearing and distance

tangents to curves from specified points

curves by specifying two elements of the curve data

geometric shapes with or without specifying size

lines from the exact endpoints of other lines

lines perpendicular or parallel to other lines

parallel lines at specified offset distances

mirror image of all or a portion of a drawing

curves tangent to one or two lines by specifying two elements of the curve data

Change the scale of all or a selected portion of a drawing

Rotate all or a portion of a drawing by a specific or unspecified amount

Copy an area or one element of a drawing to another location or to multiple locations

Extend lines to some specified locations or to some specified lengths

Trim lines to some specified location or to some specified length

Label lines and curves with bearing and distances or curve data

Label coordinates

Trace information into the computer

This is a very brief description of the most basic work that CADD can perform for civil engineers. It is like a drop of water in an ocean of power. More specific descriptions of the capabilities of CADD and other software are given in the relevant chapters.

Spreadsheets

Spreadsheets are among the most flexible software tools an engineer can use. A spreadsheet is a matrix of columns and rows. The columns are identified by letters, the rows by numbers. The intersection spaces are called *cells* and are identified by the letter and number corresponding to the location of that cell. Cells can contain values (numbers), formulas, or labels. The way of expressing the formulas differs among software products. This book uses Microsoft Excel illustrations.

One of the simplest uses of spreadsheets for civil engineers is for the preparation of cost estimates. As an example, the first column might be a list of the materials needed for construction of a particular project. The first row would label what is in each of the columns, such as (A) quantity, (B) units, (C) description, (D) unit price, (E) total. The engineer then would fill in the following:

The value for the quantity in the cell created by the first column, second row (A2)

The label for the units in the cell created by the second column, second row (B2)

The label for the description (C2). The value for the unit price (D2)

The formula for the cost of the quantity of material in row 2 is = A2 * D2 as expressed in the spreadsheet Microsoft Excel. That is, the value in cell A2 times the value in cell D2 equals the value in the cell in which the formula is expressed—in this case column E. If there is to be a total of the values in column E, column E is highlighted though a drag and drop task then the sum (Σ) button is clicked.

The beauty of a computerized spreadsheet is that once it has been set up, you can change any of the values and the corresponding total values will be changed automatically. This provides the engineer the option of using different values for different projects without having to re-create the spreadsheet.

Most spreadsheet programs offer trigonometric values, exponential operations, logic functions, and string functions. With these resources you can develop programs to solve any mathematical problems you have. There are software programs available for most civil engineering problems. However, if the problem is not complicated or has some unique elements, it may be more efficient simply to write the program yourself.

Software programs for solving particular problems are described in the chapter covering the subject.

Summary

The resources needed to design land development projects are many and varied. Many of the resources are not identified to college students and must be learned on the job. Knowing about maps and plans that may exist in the area of the development is essential. Assessors' parcel maps, USGS maps, and geographic information systems are some of the first resources needed. A site visit is essential.

Engineers need to be able to identify and locate property. Familiarity with title reports, subdivision maps, metes-and-bounds descriptions, and descriptions by townships and sections is necessary. Coordinate systems are used to establish existing and planned properties and structures exactly.

Design engineers are absolutely dependent on accurate surveys for design and construction of projects. Engineers need to understand the various kinds of surveys and their uses. Modern surveying technology and state-of-the-art computer software is changing and developing at a rapid pace. Availability of maps and plans, surveys, and computer software is changing rapidly. Engineers need to have a thorough understanding of the resources available and why each is important. The design engineer must keep abreast of the changes and improvements of civil engineering software to stay competitive.

Problems

1. Define GIS.

2. Name three ways civil engineers can use the Internet in their work.

3. What is GPS?

4. What is the purpose of zoning?

5. Explain APN 678-54-04.

6. What is the purpose of a title report?

7. How do title reports help engineers?

8. Write a metes-and-bounds description of Fig. 2.6.

9. What is the BLM?

10. What is the approximate distance between the southwest corner of Section 36, T1N, R1E MDM and the northeast corner of Section 1, T1S, R1E MDM?

11. What is the approximate distance between the northwest corner of Section 10, T3S, R2W SBM and the southeast corner of the northwest quarter of Section 10, T1N, R2E?

12. Point A has coordinates N53,000, E26,560. Point B has coordinates N56,532, E28,303. What are the latitude and departure between point A and point B?

13. For the locations in Problem 11, what is the bearing and distance between point B and point A?

14. What are the four sets of unknowns that can be solved using coordinates and trigonometry?

15. What are control surveys?

16. Define *backup*. Tell why and how it should be done.

Further Reading

American Congress on Surveying and Mapping, *Metric Practice Guide for Surveying and Mapping,* Bethesda, MD 20814, 1992.

Anon, "CE News survey finds out which firms are on-line," *Civil Engineering News,* Marietta, Georgia, May 1997, p. 12.

Brown, Curtis Maitland, and Winfield H. Eldridge, *Evidence and Procedures for Boundary Location,* 2d ed., John Wiley & Sons, New York, 1981.

Brown, Curtis M., Walter G. Robillard, and Donald A. Wilson, *Boundary Control and Legal Principals,* 3d ed., John Wiley & Sons, New York, 1986.

Davis, Raymond E., Francis S. Foote, and Joe W. Kelly, *Surveying Theory and Practice,* 6th ed., McGraw-Hill, New York, 1981.

Freedman, Alan, *The Computer Glossary,* AMACOM, New York, 1991.

"GIS Project Earns $100,000 State Grant for Innovation," *Civil Engineering News,* October 1992, p. 22.

Gookin, Dan, *DOS for Dummies,* IDG Books Worldwide, San Mateo, California, 1991.

Goubert, Didier, and Robert Newton, "Small Utility GIS," *Civil Engineering,* November 1992, pp. 69–71.

Microsoft Corporation, *Getting Results with Microsoft Office for Windows,* Redmond, Washington, 1995.

Moffitt, Francis H., and Harry Bourchard, *Surveying,* 8th ed., Harper College, New York, 1989.

U.S. Department of the Interior, *Manual of Instructions for the Survey of the Public Lands of the United States,* U.S. Government Printing Office, Washington, D.C., 1973.

Wattles, William C., *Land Survey Descriptions,* 10th ed., Gurdon H. Wattles, Orange, California, 1974.

Site Analysis

Site analysis should be done by experienced engineers with a firm grasp of the technology described in the other chapters of this book. This subject is included here only as a natural beginning in the progression of a project. This chapter will be better understood and more useful after the novice has read the rest of this book.

Before engineering design is begun, a thorough analysis of the site must be made. This analysis should consist of an engineering feasibility investigation, a preliminary design, and a cost estimate. If the project is a public works project, the analysis may be elaborate and take months or even years and involve numerous public information meetings and hearings and investigation of several alternatives. Analysis of the feasibility of a site or of alternative routes is so significant that the California Department of Transportation requires that engineers follow procedures described in a manual devoted exclusively to project development.

Ideally, analysis is done at the time the client is considering several sites. This way, an intelligent choice among alternatives, based on comparisons of development and operating costs, can be made. Only the engineering feasibility will be discussed here. The preliminary design and cost estimate will be discussed in Chap. 4.

A site analysis checklist is shown in Fig. 3.1. This, or something similar, should be used to collect information. Opposite "Responsible Jurisdiction," in the space provided for remarks, put such information as "Main Street will have to be improved to 106 ft width according to George Smith of Springfield," or "Solar access for south side of Main Street is required." Any of the items on the checklist that are not compatible with the proposed development could stop the project. Find out before thousands of dollars have been spent for fees for planning and design. Keep in mind budget constraints for the analysis. Verify with the client whether they are prepared to pay for a formal report. On a

SITE ANALYSIS CHECK LIST

Date Assignment Received   Report
                                    Required
Assessor's Map Number   Property Size
Owner: Developer:
Description of Location
Description of Environment
Description of Site
Existing Zoning   Required
                                   Zoning

	YES	NO
Planning Commission Hearing	____	____
City Council Meeting	____	____
Annexation	____	____
Parcel Map	____	____
Subdivision Map	____	____
Lot line Adjustment	____	____
Other	____	____

Responsible Jurisdiction

Streets:		
Sanitary District	____	____
Flood Control District	____	____
Water Supply District	____	____
Electricity	____	____
Gas	____	____
Communication	____	____
Cable TV	____	____
Other	____	____

Reports Required	YES	NO
E.I.R.		
Negative Declaration	____	____
Traffic	____	____
Air Pollution	____	____
Noise Pollution	____	____
Hazardous Materials	____	____
Geologic	____	____
Soils	____	____
Archaeology	____	____
Historical	____	____
	____	____

Permits and Fees

Storm Drain Connection Permit	____	____
Sanitary Sewer Connection Permit	____	____
School Impacts	____	____
Highway Encroachment	____	____
Bays and Harbors	____	____
Fish and Game	____	____
	____	____

Figure 3.1 Site analysis checklist.

small project, the client may simply want the engineer to make a site visit and draw from personal experience as to what problems there will be. This chapter describes the procedure to use for a simple, formal report. A judgment should be made as to how much material can be included in the report.

Zoning Considerations

Although zoning considerations are within the domain of planning rather than engineering, it is important to be acquainted with them. Find out how the property is zoned. If the zoning is not appropriate, find out what steps to pursue to get the zoning changed and how long they will take. Typically, zoning changes require public hearings, and hearing dates may not be available for 6 months or even a year. Find out what the appeal procedures and time frames are in case the proposed change fails.

Zoning affects all sites, but each site is unique in the particular patchwork of jurisdictions that can impose constraints on development. Since the Environmental Protection Act was passed, an Environmental Impact Report or Negative Declaration has been required on most development projects. If the site is small, it may be exempt from the report. But the exemption must be established formally through specific procedures. Some other areas of concern that may require reports, clearances, and/or special permits are flood control, traffic, school district impacts, archaeological resources, hazardous materials, historic sites, noise and air pollution, scenic impacts, solar access, landslide hazards, and earthquake faults. These are examples of just a few of the things that can affect development costs. Be alert to them.

Getting Existing Plans

The next step is to get the plans for the existing streets and utilities. Plans are usually available from the public works or engineering department of the city or county where the property is located. Water, gas, electrical, telephone, and television cable lines may be within the jurisdiction of the city or may be owned by private companies. Whether or not the site can be served by the utility companies is critical.

Storm Drains and Sanitary Sewers

If the site is in a city and sewer plans are not available there or from a special district, the sewers may have been installed before the site was annexed to the city. In that case, the plans may be at county offices. In developed areas, look at plans for the improvement of adjacent tracts. Plans for adjacent tracts will show not only the sewers installed on those sites, but previously existing facilities to which they are connected. In some cases, utilities are located in easements on private property rather than within public rights-of-way.

The designs for the construction of the storm and sanitary sewers are usually on the same plans, if not as construction drawings, then as existing or proposed lines. Get copies of plans showing all adjacent utilities. The plans collected now will be used by the project engineer to indicate inverts and to help determine horizontal locations of existing lines. Elevations on different sets of plans may have been taken from different benchmarks, so it is not the elevation of inverts that is important but rather the depth of the manhole, that is, the difference between the manhole rim and the invert.

In many areas, storm and sanitary sewerage systems are the responsibility of special districts. The boundaries of these districts seldom coincide with city limits lines. Whoever is responsible for each system will have a master plan. The master plan will show the entire network of sewers or drains within that district as well as proposed lines. The agent will know which sewers are available. In some areas, the sewage treatment plant has reached capacity. That is, as much wastewater as can be processed is already reaching the plant or has been allocated, so that no more connections can be made to the system. If this is the case, ask if plant expansion is planned and, if so, when capacity will be available. If no expansion is planned, ask if there is another sanitation district in the area that might be used, or if the sewage can be treated with septic tanks. Do not give up easily. Remember, if a way to provide a sanitation system is not found, the site cannot be developed.

Streets and Signalization

Engineers in the city or county offices will dictate street dimensions and depths of pavement sections as well as any necessary signalization. Plans of existing streets in the area, however, suggest what to use for the report and cost estimate. Signalization is an expensive item. If installation of signals might be required, check with the responsible department.

Other Utilities and Services

Ordinarily, storm drains and sanitary sewers are designed for gravity flow, so their depth and slope are critical factors. Other utilities can be designed around the sewer systems. For this reason, the exact location of the other utilities is of less concern at this point. The major concern is whether the site can be served. Remember to check on water, gas, electricity, garbage collection, telephones, and cable television.

If services will have to be extended to the site, what will the costs be? Also, check to see if there will be a requirement to put aerial electrical, telephone, and television lines underground. Usually, the information needed can be learned by calling the utility companies. Ask their representative to send a copy of their facilities in the area of the site. Though these copies usually have a disclaimer, they will provide information for the project engineer and evidence of your sources in case a problem with the location of the utility comes up later. Some utility companies will not send copies, but will allow the information to be copied in their office. If time allows, make these copies to include in the file for the job. If the representative promises to send a copy, check after a week or two as to whether it has arrived; if it has not, call the representative and ask if it has been sent.

The Site

A visit to the site is imperative. The visit should be discussed with the client. If the purchase price is still being negotiated or the project is unpopular, the investigation should be made inconspicuously—possibly without entering the property. On the other hand, if the client owns the property and it is occupied, a call ahead can clear the way.

Taking notes and photographs

Plan to take careful notes on the site. Neat, orderly notes are worth the extra effort; they are more credible and are easier to use. Take along the following items:

1. Clipboard and note pad.
2. Two small 1:1000 to 1:5000 (100- to 500-scale) site maps. A copy of the assessor's map of the parcel should be available from the county assessor's office. Use one copy for marking utilities from existing plans, topography, and notes. Use the second copy to identify snapshots as described in the following.
3. Site analysis checklist (Fig. 3.1). Write comments in the Remarks sections.
4. Different colored pencils.
5. 150 mm (6-in) scale.
6. 15 or 30 m (50- or 100-ft) tape.
7. Camera. Use a Polaroid type to be sure that the snapshot does, in fact, show what is intended.
8. Extra film. A jammed film packet can cost valuable time and/or the opportunity to get pictures that are particularly needed.
9. Business cards or other identification.

Once pictures have been taken, identify what they show. Mark each snapshot with a number. Show that number on the copy of the assessor's map, at the location where the picture was taken, and use an arrow to show the direction the camera was pointed (Fig. 3.2). On a separate sheet, list the photos with a brief description of what is shown, for example, (a) existing storm sewer outfall; (b) condition of existing pavement; or (c) oak tree (Fig. 3.3). These pictures will be invaluable throughout the planning and design phases of the job.

Recognizing significant features

With experience, the significant features of any site will be conspicuous. Look for manholes, water valves, gas valves, power and telephone lines, and storm water inlets. Verify the locations of the utilities previously copied from existing plans, or sketch utilities as they exist onto one of the assessor's maps. Use different colored pencils for the utilities that can be identified, such as, red for sanitary manholes, green for storm manholes, blue for water valves. A legend on the sketch will save repetitious notes (Fig. 3.4).

Do not assume that, because there is evidence of a utility, it is in fact available. A sewer may be too shallow or already at capacity. Other utilities may present similar problems. Walk around the boundary of the

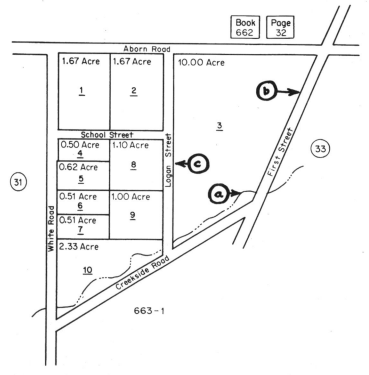

Figure 3.2   Assessor's map showing the location of objects in photographs.

Figure 3.3 Photographs taken at the site: (*a*) existing storm sewer outfall; (*b*) condition of street; (*c*) oak tree.

site. Look for other utility lines for which you may not have found plans. Look for alternative ways to service the site in case the obvious solution cannot be used. The importance of alternatives will be discussed further in other chapters of this book.

Notice the slope of the site. How much grading will have to be done? Are there rock outcroppings (Fig. 3.5)? Is the soil sandy, spongy, soggy? A soils engineer or geologist should be commissioned to prepare a preliminary report. Is the site higher in elevation than the surrounding area? If it is, find out from the water supply agency whether water is available at the elevation of the site, or whether a pump and/or storage tank will be required. If the site is lower than the surrounding area,

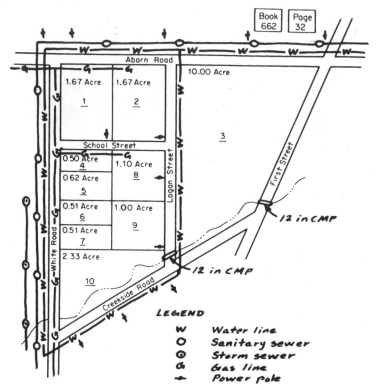

Figure 3.4 Assessor's map with utilities shown (imperial units).

Figure 3.5 Rock outcroppings.

storm water may pond there, the water table may be near the surface, or the existing sewer lines may be too shallow to be useful. How much area will be draining across the site? The storm water entering the site will have to be intercepted and prevented from flowing overland across the site.

If the drainage basin upstream of the site is large, the drainage facilities may have to be large enough to accommodate extra runoff caused by off-site flows. Is there a river or stream running through the site? If so, look for high-water marks on buildings and trees that will indicate if there has been flooding. Look at recent improvements adjacent to the waterway. There may be restrictions as to the finished floor elevation, or a flood plain corridor and dikes may have to be constructed that will limit the buildable area.

Open-space corridors in adjacent developments may indicate easements, flood plains, or *fault* zones. Notice trees. They are a valuable amenity not only for the beauty and climate control they afford but also for the structure they provide in holding slopes and possible habitat they provide. Some jurisdictions limit which trees can and cannot be cut down. Note the type and condition of buildings on the site. What appears to be a worthless shack in a rural area may in fact house the only producing water well for miles. Check it out. A standpipe (Fig. 3.6) might be part of a defunct irrigation system or a blow-off valve for the town's main water supply line. There is no way that

Figure 3.6 Standpipe.

all the possibilities can be listed. Be alert to potential trouble. Before leaving the site, review your notes. Take one last look to see if you missed anything.

Linear Projects

Linear projects such as highways, railroads, canals, power transmission lines, and major pipelines require special considerations beyond those of other kinds of projects. In urban areas, property acquisition can be the greatest single expense. Alternative routes must be compared with regard to the cost of property as well as the cost of construction. Further, when people's homes are taken to provide for a highway or airport, the residents may look to the lead agency to help them find new housing.

The Alaskan Pipeline crossed the historic migration routes of elk and reindeer, blocking passage, so the designers had to provide a way for the animals to migrate—a necessity for their survival. Engineers must also be alert to historic and anthropological sites, parks, wetlands, and natural habitats of endangered species. These national resources are assets which must be avoided if at all possible. If they cannot be avoided, mitigation will be required.

Writing the Report

When the investigation is complete, write the report. Start with an outline so the report will be clear and orderly. Have a table of contents so specific information can be found quickly. Include all pertinent information. (An example of a table of contents is given in Fig. 3.7.) Document resources. Give the names and titles of those who stated, for instance, that signalization or a 0.60 m (24-in) water main will be required, or that there will be partial reimbursement for oversized facilities.

By the time the body of the report is finished, it will be clear how the site differs from other sites. The extraordinary facilities and costs the site will require should now be known. Write a summary and put it at the beginning of the report. This way, the client will get the most important information easily. The information in the summary should not be a condensed version of the rest of the report, but rather a concise statement of selected information to describe what is unusual about the site. For instance, a part of the summary might read as follows:

> Though Clear Creek is 450 m (1480 ft) to the south, and there is no evidence of flooding in or around the site, officials at Springfield Flood Control District state that it is within the flood plain of a 100-year storm, and a flood plain zone will be required over the south 4 hectares. If flood protection can be provided, development of the site should be nearly routine.

SITE ANALYSIS
Lands of Tom and Marion Smith

Summary

I. Purpose and Limitations

II. Location and Description

III. A. General plan and zoning
    B. Evergreen development policy
    C. Impacted schools
    D. Earthquake hazards
    E. Environmental impacts
    F. Archaeological resources
    G. Noise

IV. Engineering Considerations
    A. Streets
    B. Flood protection
    C. Storm sewers
    D. Sanitary sewers
    E. Water supply
    F. Electricity
    G. Gas
    H. Telephone
    I. Cable television
    J. Grading

Conclusion

Appendix
A.   Notes
B.   Assessor's map
C.   General plan excerpt
D.   Preliminary studies, Fowler Creek benefit district map
E.   Sources

Figure 3.7 Example of table of contents.

From this, the client can go directly to the section of the report dealing with flood protection for a broader explanation of what will be required and how much it will cost. If interested, the client will also be able to find out who the officials are who stated the requirements and what a 100-year storm is.

Summary

When a developer is planning a new project, the civil engineer should be brought in as early as possible for a site evaluation. There are a number of engineering considerations that the engineer will recognize that other professionals may not. The requirements for grading, storm drains and sanitary sewers, utilities, and traffic controls will be identified. The feasibility report should be prepared by an engineer with many years of experience.

Problems

1. What is the purpose of site analysis?

2. Name three types of zones.

3. Why is it that a sanitary sewer running through the site is not sufficient evidence that the site can be served?

4. What should you check if the property is on high ground?

5. What should be checked if the property is on low ground?

6. What does a standpipe indicate?

7. Why are rock outcroppings important?

8. What should the report summary include, and where should it be located?

Further Reading

California Department of Transportation, Project Development Procedures, Sacramento, California, revised 1991.

DeChiara, Joseph, and Lee E. Koppelman, *Time-Saver Standards for Site Planning,* McGraw-Hill, New York, 1984.

National Association of Home Builders Staff, *Land Development,* 7th ed., Home Builder, Washington, D.C., 1987.

The National Association of Home Builders, *Land Development 2,* Washington, D.C., 1981.

4

Maps and Plans

The primary tools of civil engineers are maps and plans. Some types of maps are described in Chap. 2. When a new subdivision is to be created, the engineer directs the creation of the final map to provide for legal transfer of the lots within the subdivision. When a linear project is to be constructed, field surveys of the properties affected must be performed. Record of Survey maps and Assessor's Exhibits must be created so that the ownership of property can be transferred to the developer.

The engineer prepares plans to illustrate what the finished product will be so that the project can be constructed. The plans are used to prepare cost estimates so that contractors can estimate and bid on projects they wish to construct. They are also used for construction surveys so that the project can be marked with stakes in the field to inform the contractors where facilities are located horizontally and vertically.

During preparation of the plans, all of the factors that affect construction should become evident, so that problems can be addressed and solved on paper or on the computer rather than in the field during construction. The costs of solving problems during construction are much greater than solving them beforehand in the office, because contractors' personnel and equipment may be standing idle while the problem is solved.

Specifications are a companion to the plans and describe the materials and methods of construction to be used. They are described in Chap. 12.

Maps as Tools

Engineers use a variety of maps in designing improvements. Before design can begin, topographic and property boundary maps must be prepared. From information on them, preliminary maps are prepared that show the layout of the streets and lots. When preliminary site maps are prepared for private developments, the developer selects one and a *tentative map* is prepared for presentation to city council members or other committees and agencies for approval. Once approvals have been obtained, the final map (parcel map, subdivision map, or condominium map) can be prepared.

Topographic maps

A topographic map illustrates all the important physical characteristics of a site (Fig. 4.1). Anything that may affect demolition, design, or construction, and anything that may have to be protected, saved, accommodated, or connected to should be shown. Their type (concrete, asphalt, brick) and condition should be noted on the map. Some items that should be shown include the following:

Buildings

Streets and roads

Railroads, main lines, and spurs

Sidewalks

Curbs and gutters

Trees (include type and trunk diameter), streams, rock outcroppings

Elevations and contours (described in Chap. 5)

Street lights and traffic signals

Power and telephone poles

Fences and walls

Manholes, water and gas valves and meters

Transformers and telephone and communications pull boxes

Underground conduits

Wells

Underground tanks and cesspools

Easements

Flood plain zones

Earthquake fault zones

Hazardous waste sites

The agencies whose approvals will be needed may specify what symbols should be used for the various topographic features. If not, the symbols should be clear, and a legend should be included on the map.

It is important for the designer to collect all available maps and plans and to visit the site before dispatching a survey crew to collect topographic information. The engineer needs to evaluate from existing

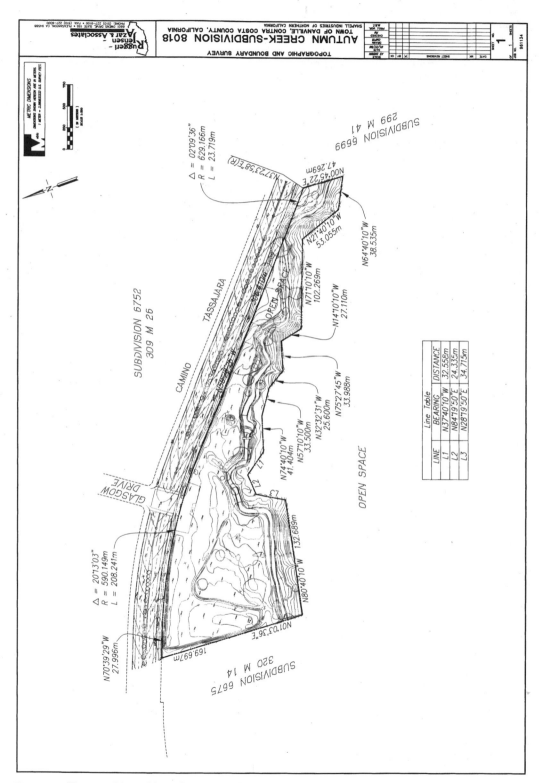

Figure 4.1 Topographic and boundary map. (*Courtesy of Rugeri, Jensen, Azar & Associates.*)

and visual information what specific data is needed from the field survey. It probably will not be apparent to the field crew if it is necessary to collect information many feet beyond the edge of the property to illustrate drainage patterns or available storm drain and sanitary sewer lines. Research of existing plans will tell the engineer what information to request.

Boundary maps

Boundary surveys should be prepared whenever improvements are to be made near property lines or when structures are to be constructed near building setback lines. If the developer makes changes within what are thought to be the property boundaries and the developer is wrong, the neighboring land owner may well demand that the improvements be removed. It is not uncommon for land owners to be mistaken about where their exact boundaries lie. The only way to determine property boundaries with certainty is with a boundary survey performed by a qualified surveyor. Even then, the neighboring land owner may hire another surveyor with a different interpretation, so that the matter will have to be settled in court. Having the boundary and topographic information on the same map is usually the best resource to use for the engineering design.

Preliminary maps

Ideally, boundary and topography maps should be complete when the preparation of preliminary maps is started. However, because timing is critical in land development, work on the preliminary map may precede the boundary and topography maps. When this happens, approximate boundaries are often plotted from the deed. Locations of items of topography that may affect the design, such as trees, should be plotted as closely to the correct location as possible.

Planners and architects often plot the initial layout of buildings or lots and streets. Their concerns are for such things as compliance with zoning, efficient use of the area, and aesthetics. The client may want the project to be a monument to his or her success, to have the ambiance of a college campus, or, more likely, to include as many lots or as much building area as physically possible. It is the engineers' or planners' task to accommodate these criteria and the physical conditions and to present various alternative preliminary maps to the client for selection.

Attention must be given to existing easements. If a neighbor, utility company, or other owner has the right to use a portion of the property for utilities or access to other properties, a portion of the client's property may have limited use. This information is stated in the title report and should be verified so that all of the constraints on the property are known.

If the site has a significant amount of slope, areas for cut or fill slopes between the lots or building pads will be necessary. The amount of space occupied by slopes must be delineated or retaining walls must be planned so their cost will be reflected in preliminary cost estimates. A good deal of money may be spent on the planning and engineering of a project only to have the project require redesign if easements and slopes are not accounted for in the early stages.

Tentative maps

When the preliminary maps are complete and one alternative has been selected, a tentative map is prepared (Fig. 4.2). The tentative map serves to show approving agencies what is planned. What is shown on a tentative map can be as simple as a proposed division of a lot into two parcels or as complicated as a new town. State and local laws and ordinances determine the circumstances under which a tentative map must be prepared and which agencies will have rights of approval. The requirement of approvals is primarily to protect public health and safety. It may be required that the map show drawings of such things as street widths; percentages of coverage of the property by buildings, streets, and landscaping; agencies supplying utilities; number of people to occupy the site; and amount of traffic generated by the project.

Approving agencies can require whatever information they deem appropriate. Usually the tentative map will be approved subject to certain conditions. The Conditions of Approval will have to be satisfied. When the Conditions of Approval are available, it is good practice to copy them and attach them to the tentative map. The project engineer can then take the tentative map and Conditions of Approval and prepare the subdivision map and construction plans.

Record of Survey maps

When the survey is performed relating to land boundaries or property lines, a Record of Survey map may be required (Fig. 2.12). In California, the map must be filed with the office of the county surveyor for the county in which the property lies. The Professional Land Surveyor's Act describes what conditions require a Record of Survey map and dictate what the map must show.

Parcel maps

A parcel map (Fig. 4.3) shows exact legal location and description of plots of land. The map is used in place of metes-and-bounds descriptions to divide property into two or more parcels. In California, parcel maps may show no more than five parcels. A subdivision map is required when more than five parcels are created.

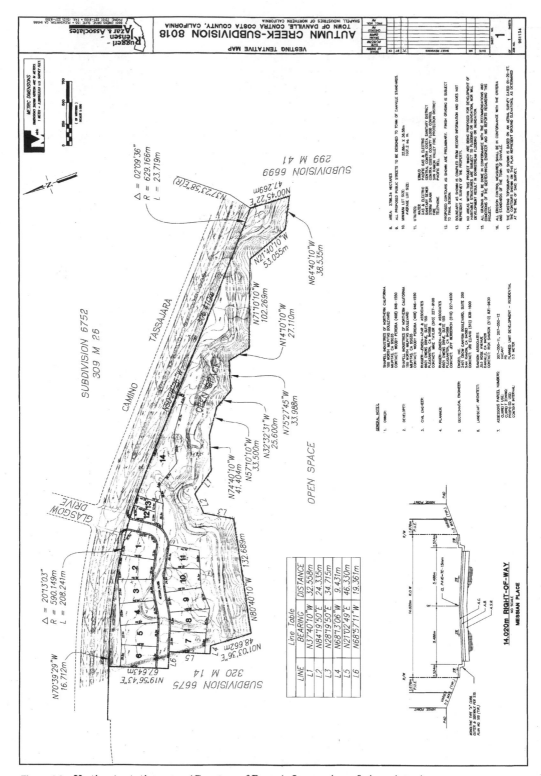

Figure 4.2 Vesting tentative map. (*Courtesy of Rugeri, Jensen, Azar & Associates.*)

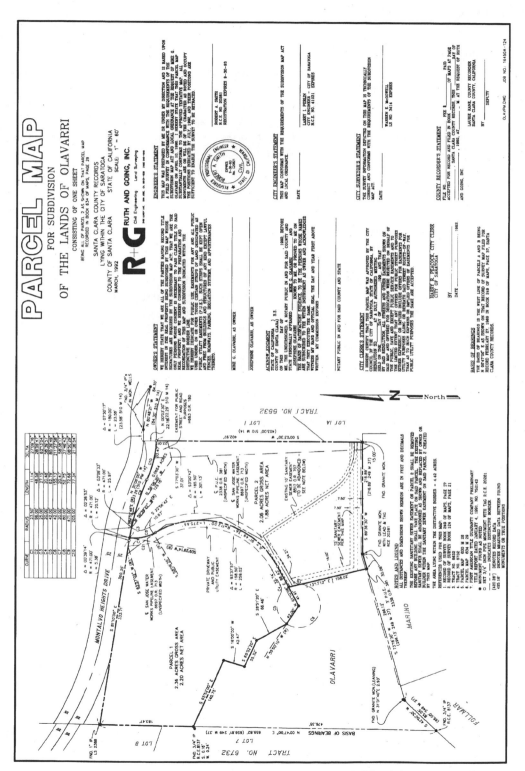

Figure 4.3 Parcel map. (*Courtesy of Ruth and Going, Inc., San Jose, California.*)

Lot line adjustments

Some places there are existing adjacent lots that due to the sizes of the two areas or the configuration of the lines between them, one or both do not satisfy criteria as buildable lots. When the lot line between them can be moved or reshaped to make both lots meet criteria a *lot line adjustment* can be made. The process for adjusting the line may make it more cost-effective than creating a new parcel map.

Subdivision and condominium maps

A subdivision map is a legal document for creating new lots on the ground in two-dimensional space. It is sometimes called a tract map (Fig. 4.4). Subdivision maps that create more than five parcels may be called tract maps, and the property may be assigned a tract number. The assignment of a tract number facilitates referencing. Condominium maps create legal documents for creating three-dimensional lots or airspace within a building. They provide ownership of an apartment or office unit (airspace) with a proportional undivided ownership of the underlying real property.

The owner of a townhouse unit holds exclusive right to the underlying land, and a subdivision or tract map is used to define ownership. An undivided ownership to common areas may be held by a townhouse owner. These maps become legal documents when they have been signed by the land owner, the surveyor or qualified civil engineer, and the consenting agencies and filed for record with the county recorder. From that point on, property can be bought and sold by reference to that map.

Final maps

When the surveying and engineering are done, all of the Conditions of Approval have been met, and the drafting is complete for a subdivision, parcel, or condominium map, it is known as a final map. Once all necessary parties have signed the map, it can be filed with the county recorder. It is then a legal document and can be used to transfer property ownership.

ALTA maps

The members of the American Land Title Association (ALTA) in association with the American Congress on Surveying and Mapping (ACSM) have developed strict standards for surveying to be used to establish existing topographic conditions on property. The reason for this is that the standards in one part of the country may differ from those in another part of the country and it may be necessary to have very exacting information to provide to lenders before loaning money for pur-

SUBDIVISION 8018
AUTUMN CREEK

BEING PORTIONS OF LOT 46 OF SUBDIVISION 6976, FILED IN BOOK 320 OF MAPS AT PAGE 14, LOT 1 OF SUBDIVISION 6518, FILED IN BOOK 309 OF MAPS AT PAGE 8, AND PARCEL A OF SUBDIVISION 7164, FILED IN BOOK 350 OF MAPS AT PAGE 48, CONTRA COSTA COUNTY RECORDS

TOWN OF DANVILLE
CONTRA COSTA COUNTY   CALIFORNIA

RUGGERI–JENSEN–AZAR & ASSOCIATES
CIVIL ENGINEERS, PLANNERS, SURVEYORS
PLEASANTON, CALIFORNIA
JANUARY 1998

OWNER'S STATEMENT

THE UNDERSIGNED, BEING THE PARTY HAVING A RECORD TITLE INTEREST IN THE LANDS DELINEATED AND EMBRACED WITHIN THE HEAVY BLACK LINES UPON THIS MAP, ENTITLED "SUBDIVISION 8018, AUTUMN CREEK", DO HEREBY CONSENT TO THE MAKING AND RECORDATION OF THE SAME; AND DO HEREBY DEDICATE TO THE PUBLIC FOR PUBLIC USE THOSE PORTIONS OF SAID LANDS DESIGNATED ON SAID MAP AS MESSIAN PLACE FOR PUBLIC STREET PURPOSES.

THE AREAS DESIGNATED AS PARCELS A, B, C, AND D ARE FOR USE BY THE HOMEOWNERS ASSOCIATION.

THE AREA DESIGNATED "PUBLIC UTILITY EASEMENT" OR "P.U.E." IS HEREBY DEDICATED TO THE PUBLIC FOR PUBLIC USE FOR "PUBLIC UTILITY PURPOSES" INCLUDING CONSTRUCTION, ACCESS OR MAINTENANCE OF WORKS, IMPROVEMENTS AND STRUCTURES, AND THE CLEARING OF OBSTRUCTIONS AND VEGETATION.

THE AREA DESIGNATED "STORM DRAIN EASEMENT" OR "S.D.E." IS DEDICATED TO THE TOWN OF DANVILLE OR ITS DESIGNEE AND TO THE PUBLIC FOR PUBLIC USE FOR STORM, FLOOD AND SURFACE WATER DRAINAGE, INCLUDING CONSTRUCTION, ACCESS FOR MAINTENANCE OF WORKS, IMPROVEMENTS AND STRUCTURES, WHETHER COVERED OR OPEN, AND FOR THE CLEARING OF OBSTRUCTIONS AND VEGETATION.

THE AREA DESIGNATED AS "PRIVATE STORM DRAIN EASEMENT" OR "P.S.D.E." IS FOR THE BENEFIT OF LOTS 12, 13 AND PARCEL E FOR STORM, FLOOD AND SURFACE WATER DRAINAGE, INCLUDING CONSTRUCTION, ACCESS FOR MAINTENANCE OF WORKS, IMPROVEMENTS AND STRUCTURES, WHETHER COVERED OR OPEN, AND FOR THE CLEARING OF OBSTRUCTIONS AND VEGETATION.

THE AREA DESIGNATED AS "SANITARY SEWER EASEMENT" OR "S.S.E." IS OFFERED FOR DEDICATION TO THE CENTRAL CONTRA COSTA SANITARY DISTRICT (CCCSD) OR ITS DESIGNEE IN GROSS AS AN EXCLUSIVE SUBSURFACE EASEMENT AND NON-EXCLUSIVE SURFACE EASEMENT FOR SANITARY SEWER PURPOSES INCLUDING CONSTRUCTION, ACCESS OR MAINTENANCE OF WORKS, IMPROVEMENTS, AND STRUCTURES, AND THE CLEARING OF OBSTRUCTIONS AND VEGETATION. NO BUILDING OR STRUCTURE MAY BE PLACED ON SAID EASEMENT NOR SHALL ANYTHING BE DONE THEREON WHICH MAY INTERFERE WITH CCCSD'S FULL ENJOYMENT OF SAID EASEMENT.

THE AREA DESIGNATED AS "TRAIL EASEMENT" IS DEDICATED TO THE TOWN OF DANVILLE FOR PUBLIC ACCESS OVER TRAILS DESIGNATED BY THE TOWN OF DANVILLE AS BEING FOR THE PUBLIC USE. (THIS EASEMENT RIGHT DOES NOT INCLUDE GENERAL PUBLIC ACCESS TO ANY AREA OTHER THAN DESIGNATED PUBLIC TRAILS).

THE AREA DESIGNATED AS "SCENIC EASEMENT" IS DEDICATED TO THE TOWN OF DANVILLE AS AN AREA TO REMAIN IN ITS NATURAL AND SCENIC STATE FOR THE BENEFIT AND ENJOYMENT OF THE OWNERS OF THIS SUBDIVISION AND THE GENERAL PUBLIC. THE TOWN OF DANVILLE'S EASEMENT RIGHT IS TO DENY ANY AND ALL GRADING, LANDSCAPING, DEVELOPMENT AND STRUCTURES WITHIN THIS AREA, EXCLUSIVE OF ANY OTHER IDENTIFIED EASEMENT RIGHTS (THIS EASEMENT RIGHT DOES NOT INCLUDE GENERAL PUBLIC RIGHT OF ENTRY).

THE AREA DESIGNATED AS "PARCEL E" OR "OSBORN WAY" IS NOT DEDICATED FOR PUBLIC USE BUT IS FOR THE BENEFIT OF LOTS 12, 13 AND 14 OF THIS MAP FOR PRIVATE ACCESS.

THE AREA DESIGNATED AS "LANDSCAPE MAINTENANCE EASEMENT" IS DEDICATED TO THE TOWN OF DANVILLE, OR ITS DESIGNEE, FOR LANDSCAPE MAINTENANCE PURPOSES.

THE AREA DESIGNATED AS "PRIVATE LANDSCAPE EASEMENT" IS FOR USE BY THE HOMEOWNERS ASSOCIATION FOR LANDSCAPE MAINTENANCE PURPOSES.

THE AREA DESIGNATED AS "WALL MAINTENANCE EASEMENT" IS DEDICATED TO THE TOWN OF DANVILLE FOR WALL MAINTENANCE PURPOSES.

THE AREA DESIGNATED AS "EMERGENCY VEHICLE ACCESS EASEMENT" OR "EVAE" IS HEREBY DEDICATED TO THE TOWN OF DANVILLE, OR ITS DESIGNEE, FOR PUBLIC USE FOR EMERGENCY VEHICULAR ACCESS.

THE AREA DESIGNATED AS "NON DEVELOPMENT EASEMENT" IS DEDICATED TO THE TOWN OF DANVILLE AS A NON DEVELOPMENT AREA EXCEPT FOR ROADWAY AND UNDERGROUND UTILITIES.

THE AREA DESIGNATED AS "EBMUD" IS DEDICATED TO EAST BAY MUNICIPAL UTILITY DISTRICT AS A PERPETUAL EASEMENT FOR THE PURPOSE OF CONSTRUCTING, REPLACING, MAINTAINING, OPERATING AND USING, AS THE GRANTEE MAY SEE FIT, FOR THE TRANSMISSION AND DISTRIBUTION OF WATER, A PIPE OR PIPELINES AND ALL NECESSARY FIXTURES INCLUDING UNDERGROUND TELEMETRY AND ELECTRICAL CABLES OR APPURTENANCES THERETO, IN, UNDER, ALONG AND ACROSS SAID EASEMENT, TOGETHER WITH THE RIGHT OF INGRESS TO AND EGRESS FROM SAID EASEMENT AND EVERY PART THEREOF. THE GRANTOR AND THE GRANTOR'S HEIRS, SUCCESSORS, OR ASSIGNS SHALL NOT PLACE OR PERMIT TO BE PLACED ON SAID EASEMENT ANY BUILDING OR STRUCTURE, NOR DO NOR ALLOW TO BE DONE ANYTHING WHICH MAY INTERFERE WITH THE FULL ENJOYMENT BY THE GRANTEE.

THE ABOVE PARAGRAPH NOTWITHSTANDING GRANTOR RESERVES THE RIGHT TO LANDSCAPE THE EASEMENT AREA IN A MANNER CONSISTENT WITH THE GRANTEE'S USE; HOWEVER, SUCH USE BY GRANTOR SHALL NOT INCLUDE THE PLANTING OF TREES NOR A CHANGE IN THE EXISTING SURFACE ELEVATION (GRADE) OF THE EASEMENT AREA BY MORE THAN ONE (1) FOOT WITHOUT FIRST HAVING PRIOR WRITTEN CONSENT OF THE GRANTEE.

THE MAP SHOWS ALL EASEMENTS ON THE PREMISES OR OF RECORD.

OWNER: STANDARD PACIFIC CORP., A DELAWARE CORPORATION

DATE: _____   BY: _____   MICHAEL CORTNEY, PRESIDENT

ACKNOWLEDGEMENT

STATE OF CALIFORNIA)
COUNTY OF CONTRA COSTA)

ON _____, 1998, BEFORE ME, _____, A NOTARY PUBLIC IN AND FOR SAID COUNTY AND STATE, PERSONALLY APPEARED _____

PERSONALLY KNOWN TO ME (OR PROVED TO ME ON THE BASIS OF SATISFACTORY EVIDENCE) TO BE THE PERSON WHOSE NAME IS SUBSCRIBED TO THE FOREGOING STATEMENT AND ACKNOWLEDGED TO ME THAT HE EXECUTED THE SAME IN HIS AUTHORIZED CAPACITY, AND THAT BY HIS SIGNATURE ON THE STATEMENT THE PERSON OR THE ENTITY UPON BEHALF OF WHICH THE PERSON ACTED, EXECUTED THE STATEMENT.

WITNESS MY HAND,

_____ (SIGNATURE)

_____ (PRINT)

MY COMMISSION EXPIRES: _____

COUNTY OF PRINCIPAL PLACE OF BUSINESS: _____

CITY CLERK'S STATEMENT

I, ROCHELLE FLOTTEN, DEPUTY CITY CLERK AND EX-OFFICIO CLERK OF THE TOWN COUNCIL OF THE TOWN OF DANVILLE, COUNTY OF CONTRA COSTA, STATE OF CALIFORNIA, DO HEREBY STATE THAT THE HEREIN EMBODIED FINAL MAP ENTITLED "SUBDIVISION 8018, AUTUMN CREEK" WAS PRESENTED TO SAID COUNCIL AS PROVIDED BY LAW, AT A REGULAR MEETING THEREOF HELD ON THE _____ DAY OF _____, 1998, AND THAT SAID COUNCIL DID THEREUPON BY RESOLUTION APPROVE SAID MAP AND DID ACCEPT, SUBJECT TO THE ACCEPTANCE OF PUBLIC IMPROVEMENT, ON BEHALF OF THE PUBLIC ANY OF THE ROADS OR EASEMENTS SHOWN THEREON AS DEDICATED TO PUBLIC USE AND DOES HEREBY ABANDON THOSE EASEMENTS IDENTIFIED HEREIN PER SECTION 66499.20 1/2 OF THE SUBDIVISION MAP ACT.

I FURTHER STATE THAT ALL BONDS AS REQUIRED BY LAW TO ACCOMPANY THIS FINAL MAP HAVE BEEN APPROVED BY THE TOWN COUNCIL OF THE TOWN OF DANVILLE AND ARE FILED IN MY OFFICE.

IN WITNESS WHEREOF, I HAVE HEREUNTO SET MY HAND THIS _____ DAY OF _____, 1998.

ROCHELLE FLOTTEN
DEPUTY CITY CLERK AND EX-OFFICIO CLERK OF THE TOWN OF DANVILLE, COUNTY OF CONTRA COSTA STATE OF CALIFORNIA

RECORDER'S STATEMENT

THE MAP ENTITLED "SUBDIVISION 8018, AUTUMN CREEK" IS HEREBY ACCEPTED FOR RECORDATION, SHOWING A CLEAR TITLE AS PER LETTER OF TITLE MADE BY FIRST AMERICAN TITLE GUARANTEE COMPANY, DATED THE _____ DAY OF _____, 1998, AND AFTER EXAMINING THE SAME, I DEEM THAT SAID MAP COMPLIES IN ALL RESPECTS WITH THE PROVISIONS OF STATE LAWS AND LOCAL ORDINANCES GOVERNING THE FILING OF SUBDIVISION MAPS.

FILED AT THE REQUEST OF FIRST AMERICAN TITLE GUARANTEE COMPANY AT _____ MINUTES PAST _____ M., ON THE _____ DAY OF _____, 1998, IN BOOK _____ OF MAPS AT PAGE _____, IN THE OFFICE OF THE COUNTY RECORDER OF THE COUNTY OF CONTRA COSTA, STATE OF CALIFORNIA.

STEPHEN L. WEIR
COUNTY RECORDER

BY: _____
DEPUTY COUNTY RECORDER

JOB NO. 960154.20 A   SHEET 1 OF 5 SHEETS

Figure 4.4 Tract map. (*Courtesy of Rugeri, Jensen, Azar & Associates.*)

SUBDIVISION 8018
AUTUMN CREEK

BEING PORTIONS OF LOT 46 OF SUBDIVISION 6678, FILED IN BOOK 320 OF MAPS AT PAGE 14, LOT 1, OF SUBDIVISION 6516, FILED IN BOOK 309 OF MAPS AT PAGE 3, AND PARCEL A OF SUBDIVISION 7104, FILED IN BOOK 350 OF MAPS AT PAGE 48, CONTRA COSTA COUNTY RECORDS

TOWN OF DANVILLE
CONTRA COSTA COUNTY   CALIFORNIA

RUGGERI-JENSEN-AZAR & ASSOCIATES
CIVIL ENGINEERS, PLANNERS, SURVEYORS
PLEASANTON, CALIFORNIA
JANUARY 1998

ENGINEER'S STATEMENT

I CERTIFY THAT THIS MAP WAS PREPARED FROM A SURVEY MADE BY ME OR UNDER MY DIRECTION IN APRIL 1997, THAT THE SURVEY IS TRUE AND COMPLETE AS SHOWN, AND THAT THE MONUMENTS OF THE CHARACTER SHOWN ON THE FINAL MAP WILL BE SET IN SUCH POSITIONS ON OR BEFORE JANUARY 2000 AND WILL BE SUFFICIENT TO ENABLE THE SURVEY TO BE RETRACED.

THE SUBDIVISION CONTAINS 16.118 HECTARES MORE OR LESS, ALL PORTIONS OF THIS MAP LIE WITHIN THE INCORPORATED AREA OF THE TOWN OF DANVILLE.

ALL BEARINGS OF THIS MAP ARE BASED ON THE CALIFORNIA COORDINATE SYSTEM ZONE III, CCS27.

DATE:_____

PIERO P. RUGGERI, R.C.E. NO. 25281
REGISTRATION EXPIRES: DECEMBER 31, 2001

CITY ENGINEER'S STATEMENT

I, STEVEN C. LAKE, CITY ENGINEER OF THE TOWN OF DANVILLE, COUNTY OF CONTRA COSTA, STATE OF CALIFORNIA, HEREBY STATE THAT I HAVE EXAMINED THIS MAP ENTITLED "SUBDIVISION 8018, AUTUMN CREEK", THAT SAID SUBDIVISION AS SHOWN IS SUBSTANTIALLY THE SAME AS IT APPEARED ON THE TENTATIVE MAP AND APPROVED ALTERATIONS THEREOF, THAT ALL PROVISIONS OF STATE LAWS AND LOCAL ORDINANCES GOVERNING THE FILING OF SUBDIVISION MAPS HAVE BEEN COMPLIED WITH, AND THAT I AM SATISFIED THAT THE SAME IS TECHNICALLY CORRECT.

DATED:_____

STEVEN C. LAKE, R.C.E. 31870
CITY ENGINEER
TOWN OF DANVILLE
STATE OF CALIFORNIA
REGISTRATION EXPIRES: DECEMBER 31, 2000

STATEMENT OF PLANNING COMMISSION

I HEREBY CERTIFY THAT THE PLANNING COMMISSION OF THE TOWN OF DANVILLE, CONTRA COSTA COUNTY, STATE OF CALIFORNIA, HAS APPROVED THE TENTATIVE MAP OF THIS SUBDIVISION, UPON WHICH THIS FINAL MAP IS BASED.

DATE:_____ 1998.

KEVIN J. GAILEY
CHIEF OF PLANNING
TOWN OF DANVILLE
COUNTY OF CONTRA COSTA
STATE OF CALIFORNIA

BUILDING INSPECTOR'S STATEMENT

A PRELIMINARY SOIL INVESTIGATION REPORT PREPARED BY ENGEO INCORPORATED, DATED DECEMBER 5, 1991, REPORT NO. N1-2092-W15, HAS BEEN RECEIVED AND APPROVED. THE REPORT IS ON FILE IN THE TOWN OF DANVILLE BUILDING INSPECTION DIVISION, DANVILLE, CALIFORNIA.

DATE:_____ 1998.

KEVIN J. GAILEY
CHIEF OF PLANNING
TOWN OF DANVILLE
COUNTY OF CONTRA COSTA
STATE OF CALIFORNIA

CLERK OF THE BOARD OF SUPERVISORS CERTIFICATE

STATE OF CALIFORNIA   )
) SS.
COUNTY OF CONTRA COSTA   )

I, PHIL BATCHELOR, CLERK OF THE BOARD OF SUPERVISORS AND COUNTY ADMINISTRATOR OF THE COUNTY OF CONTRA COSTA, STATE OF CALIFORNIA, DO HEREBY CERTIFY, AS CHECKED BELOW THAT:

☐ A TAX BOND ASSURING THE PAYMENT OF ALL TAXES WHICH ARE NOW A LIEN, BUT NOT YET PAYABLE, HAS BEEN RECEIVED AND FILED WITH THE BOARD OF SUPERVISORS OF CONTRA COSTA COUNTY, STATE OF CALIFORNIA.

☐ ALL TAXES DUE HAVE BEEN PAID, AS CERTIFIED BY THE COUNTY REDEMPTION OFFICER.

DATED:_____

PHIL BATCHELOR
CLERK OF THE BOARD OF SUPERVISORS
AND COUNTY ADMINISTRATOR

BY:_____
        DEPUTY CLERK

JOB NO. 961134.20 A

Figure 4.4   (*Continued*)

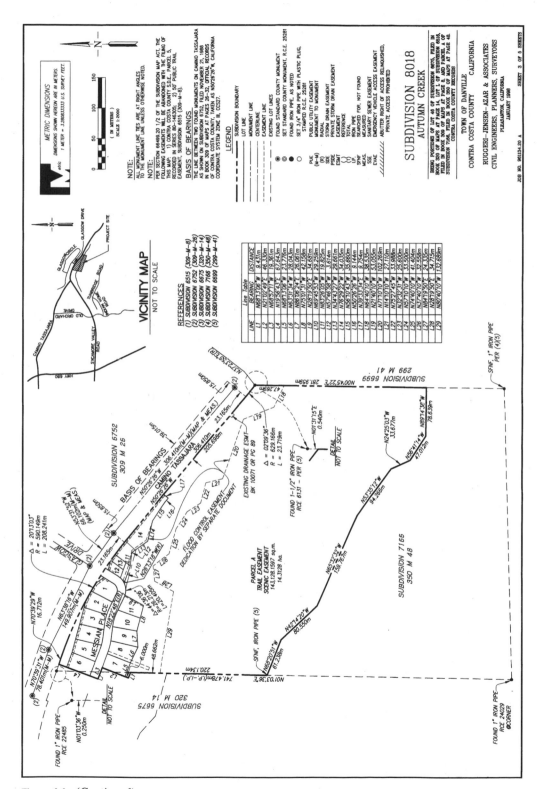

Figure 4.4 *(Continued)*

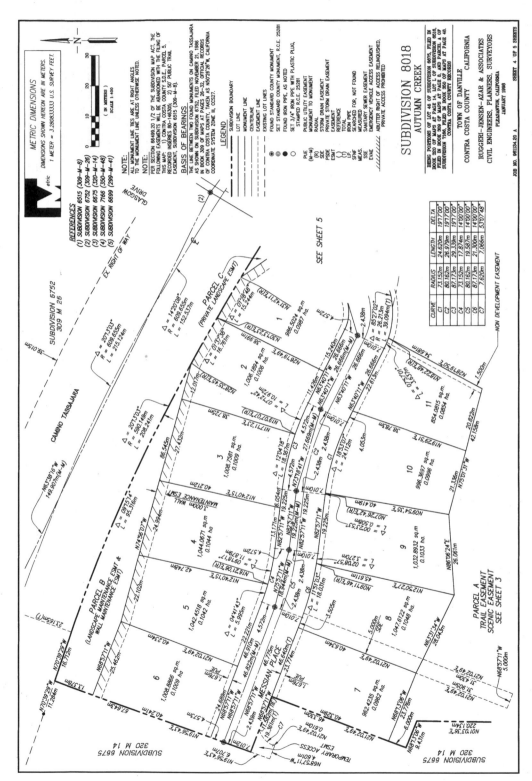

Figure 4.4 *(Continued)*

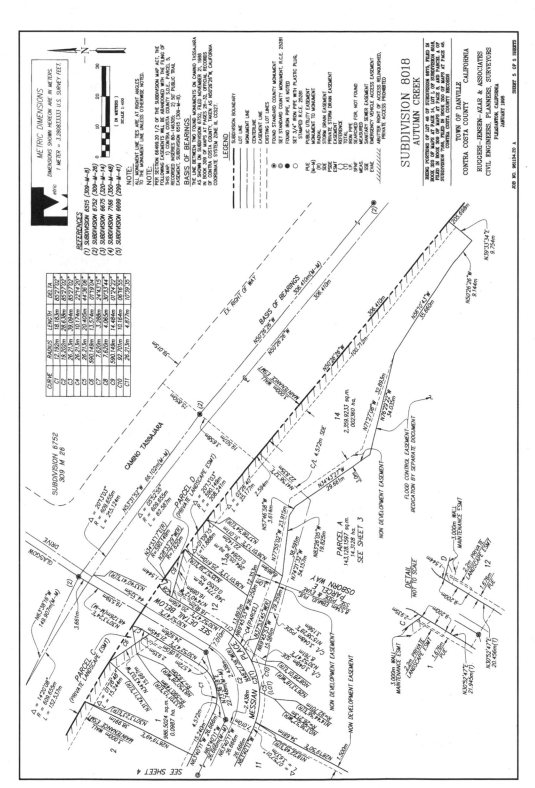

Figure 4.4 (*Continued*)

chases of land. For that reason, engineers are sometimes asked to provide topographic maps that meet ALTA standards (ALTA maps) before funds can be made available for design.

Alternative route maps

The best route between two points is seldom straight because of constraints imposed by the terrain. Even on flat deserts or plains, curves are sometimes built into highway routes so that drivers will not become hypnotized or simply bored and fall asleep. Time spent preparing and investigating alternative routes is valuable. Alternative route plans are usually shown on aerial maps or contour (topographic) maps. The route that may be the obvious choice relative to the physical (engineering) conditions may be the worst choice politically. The best choice politically may endanger the environment or have restrictive costs. Each alternative will have its advantages and disadvantages. The alternative route maps are used for comparative studies and for illustrations at public hearings. Their use allows for an informed decision among alternatives.

Elements of Plans

The term *plan* can refer to anything from saying "let's eat" to volumes of books and drawings describing methods and procedures to accomplish some result. In engineering and allied fields, plans usually refer to drawings made to scale showing horizontal, vertical, and cross-sectional views of tangible objects to be built. Plans usually include written descriptions of construction materials and procedures (specifications), artists' renderings, or any other device to help make clear what is to be accomplished.

The plan view

The plan view shows significant information, such as existing and proposed centerlines, edges of traveled way, shoulders, face of curb lines, and property lines as viewed from above (Fig. 4.5). They should also show existing and proposed power poles, storm water inlets, storm drain and sanitary sewer lines, water supply lines, utilities lines, trees, and other objects that might affect the design.

Profiles

The view from the side showing the grades, vertical curves, and appurtenant structures of any linear aspect of a project is called a *profile*. When the profile is of a street, the profile line can be located at any convenient reference location on the cross section, as long as it is clearly labeled and identified. The reference line is on the centerline for most streets. Typically, the sanitary sewer, storm drain, and other utility line

Plan view

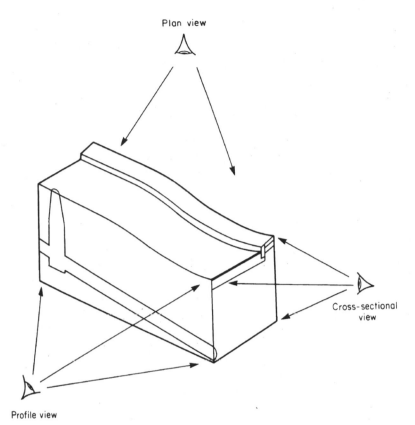

Cross-sectional
view

Profile view

Figure 4.5   Plan, profile, and cross-sectional views.

profiles are included. Though there is some horizontal distance between the various profiles, they are projected onto the same vertical plane.

Cross sections

Cross-sectional views show what is seen when an object is cut across with a vertical plane. On linear projects, the cross sections are usually made at right (90°) angles to the centerline. Cross sections are used extensively in all areas of engineering and architecture. There are several examples of cross sections throughout this book. Geometric cross sections (see Fig. 7.18) show only dimensions and slopes; structural cross sections (see Fig. 7.19) show types and depths of material to be used.

When earthwork is to be calculated, cross-sectional views are necessary if the end area formula is to be used (see Chap. 6). Cross sections are necessary on grading plans to portray how the finished product will look. They are also helpful when the natural ground is sloped and the *catch points* (see Fig. 6.8) at tops and toes of slopes are not apparent from the plan view.

Details

Most sets of plans include some detail drawings. These will be standard details established by the jurisdiction involved or may be structures that vary from the standard plans of typical structures approved for use within the jurisdiction. Detail drawings also show situations too small to be clear at the scale used for the surrounding area. The details may be shown on the cover sheet, on the sheet where the object is located, or on special detail sheets, depending on the number of details used and what will be most clear.

The detail can be a plan, profile, or cross-sectional view, or any other type of drawing that will assist in making the intent clear. Some details are dimensioned in meters or millimeters.

Title sheets

The title sheet (Fig. 4.6) provides general information about the project. It should include a title, general description of the project, and a location or vicinity map. It may include an index map of sheets or a table of contents of sheets. There are lines for the signature of the various approving agents and the design engineer and the engineering stamps of each. The approving agency will probably require a list of items to be shown on the cover sheet.

Index sheet

On large or complicated projects it may be expeditious to provide an index sheet. This will show a map of the project with outlines of interior sheets and their sheet numbers, so that those using the plans can easily find the sheet they need.

Plan sheets

Land development plans are usually prepared on sheets 24×36 in or 22×34 in, although any size can be used. The 22×34 in size is popular because when it is reduced by 50 percent, it becomes 11×17 in, which works well in reports and is easily handled. At this writing, sheet sizes have not been changed to metric units. In some cases, a larger size is used so that the civil site plans conform to other sheets in a set of plans prepared by an architect for a building or buildings.

For a very simple project, all of the necessary information may be shown on a single sheet, but in most cases several sheets are needed.

The interior sheets of a set of plans are often the plan and profile sheets (Fig. 4.7). These sheets most often are divided lengthwise. The top half of the sheet is for the drawings of the plan view of the streets. The bottom half has a grid for drawing the profile view of the streets and underground utilities. This may be reversed, with the profile on top and the plan view on the bottom. On site plans and complicated

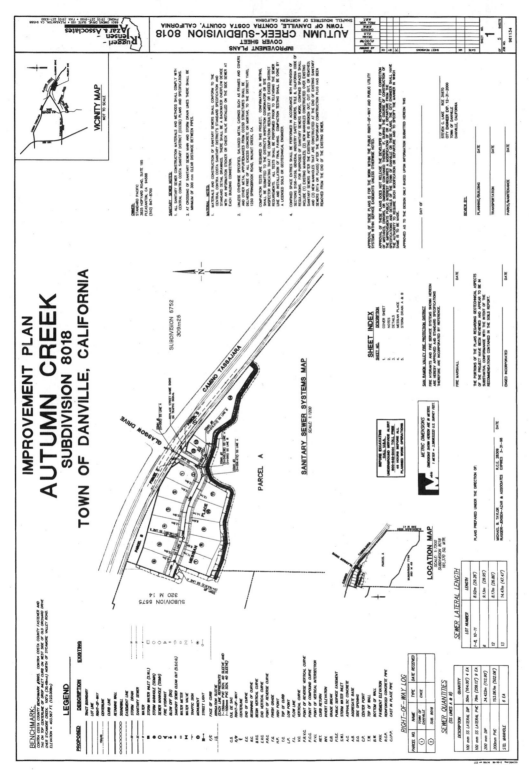

Figure 4.6 Title sheet. (*Courtesy of Rugeri, Jensen, Azar & Associates.*)

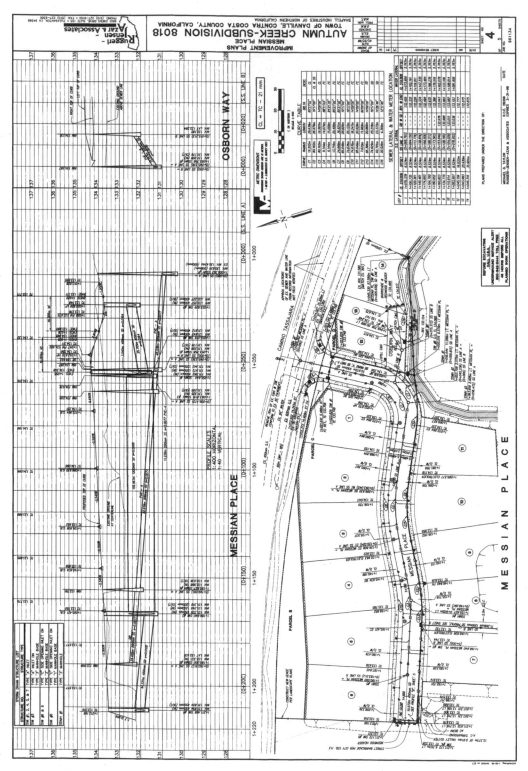

Figure 4.7 Plan and profile sheet. (*Courtesy of Rugeri, Jensen, Azar & Associates.*)

78

projects such as interchanges, the plans and profiles may be shown on separate sheets.

Types of Plans

Some of the types of plans used in land development will be briefly described in this section. It is important to remember that the nomenclature may differ from one city or state to another. For instance, the plans referred to in one location as *construction plans* may be called *improvement plans* in another locale. Do not let that distract you: the purpose and character of the plans will be the same.

The format for including the material on the various types of plans will vary among locales, engineering firms, clients, times, and jobs. All the necessary information for a simple job might be included on a single sheet with a title such as Site Plan. A large, complicated job might have several sets of plans, each containing 100 or more sheets. The number of sheets is determined by how many are needed to show clearly what is proposed and is limited only by how many sheets make the set unwieldy.

Master plans

Each project is just one piece of a giant puzzle. That piece must be made to fit the whole puzzle—the master plan. On large projects it may be helpful to have master plans for storm drainage, sanitary sewers, and water supply lines. Public agencies that are responsible for providing or regulating facilities view the storm drainage, sanitary sewerage, water supply, and traffic plans differently than engineers involved in single projects. The individual projects must fit into the larger master plan of the entire district. Therefore, the master plan of each affected agency must be consulted before work on the site plans begins.

Demolition plans

When construction begins, demolition plans will be the first to be used in the field. The purpose of demolition plans is to show what is to be saved and what is to be demolished or abandoned so contractors can clear the way for the new construction. The demolition plan is made on a reproducible copy of the topography map. Special care should be taken to verify that all underground utilities or other facilities are shown. A job can be shut down because a conduit is uncovered that was not shown on the plans. Work cannot continue until it is determined what the conduit carries.

There can be serious consequences to ripping up a 100 mm (4-in) high-pressure gas line or even a water main. It can be a matter of life or death for workers or others in the vicinity. When the presence of a large, high-pressure gas main is known, it must be marked with a highly visible note such as the word DANGER spelled out in bold letters followed by a boldly printed note of what the danger is.

The limits of demolition should be clear, and equipment should be kept a safe distance from fences and walls owned by neighbors. If excavation is required, a soils engineer should be consulted about what precautions must be taken to protect adjacent structures. Where a portion of sidewalk or other concrete area is to be preserved, there should be a note to "saw-cut or break at a joint" to ensure a professional, finished product. When some, but not all, trees or other amenities are to be saved, they must be marked in the field with an unmistakable message such as a fence erected clearly for the purpose of protecting that object. When the site is stripped of obstacles and the rubble has been cleared away, the site is ready for grading.

Grading plans

Before concrete and pipes can be laid and structures built, the site must be compacted to provide a solid, stable base. The surrounding area must be sculptured to slope away from structures to provide drainage, and parking lots and streets must be shaped and sloped to direct overland flow toward drainage facilities. The grading plan shows elevations at appropriate locations. The survey crew marks the horizontal locations with 2×2 wooden hubs and wooden stakes. The stake identifies the elevation and the amount of cut or fill required above or below the top of the hub.

Construction/improvement plans

Once the grading is finished, work on the structures can begin. The construction plans show the horizontal and vertical locations of the structures as well as cross sections and whatever other details and instructions may be necessary to build bridges, streets, drainage facilities, sewers, water supply lines, and other utilities. The improvement plans may be made up of several sets of plans covering the same area but different aspects of the construction. Some of the sets that may be needed within the improvement plans are as follows:

Cover sheet

Index sheet

Typical cross sections showing geometric and structural cross sections

Layout plans showing in plan view the location of surface improvements such as highways, roads, parking areas, and buildings

Profile plans showing the profiles of streets or highways

Grading plans

Contour grading plans

Bridge plans

Construction details

Drainage plans showing the horizontal layout of drainage structures for removal of surface drainage

Drainage profiles

Subsurface drainage plans showing the layout of drainage structures to remove subsurface water

Subsurface drainage profiles

Drainage details

Sanitary sewer plans

Sanitary sewer profiles

Sanitary sewer details

Street lighting plans

Traffic signal plans

Traffic striping/delineation plans

Detour plans

Landscape plans

Landscape irrigation plans

Landscape details

Site plans for complex commercial, industrial, or multifamily residential developments can also include the following plans:

Architectural

Structural

Mechanical

Electrical

Utility

Fire Safety

Handicap access

The number and types of plans, the sheet sizes, and the scale to be used are decisions that the engineer should make after careful consideration of the complexity of the job. When in doubt, err on the side of more sheets.

Traffic plans

Traffic plans illustrate the signalization, signing, lighting, striping, and guard rails of highways, roads, and streets for a project. When they are drawn for renovation of existing streets or roads, they may be drawn in ink on Mylars of aerial photographs. When the plans are for new streets, they may be drawn on copies of base maps for construction plans. A *base map* is a drawing which shows little other than the lines delineating properties and streets. The plan is copied and used for different types of plans of the same area. Specialized information is then added to the different copies.

Traffic plans may or may not be included in the plan set, depending on the complexity of the project and the requirements of the agency having responsibility for traffic control and safety. The plans can include traffic flow diagrams and signalization plans as well as the locations, sizes, and types of striping and other pavement markings. A civil engineer specializing in traffic engineering should be responsible for the preparation of traffic engineering plans.

Landscaping plans

Landscaping plans are required on most projects. Ordinarily the landscaping is designed by a landscape architect. The plans show the location and types of the trees and plants to be used and various other information to ensure the success of the planting. A plan for irrigation and lighting is usually included.

Utility trench plans

Power and communication lines are now installed underground in trenches rather than overhead on poles. As a result, most projects will include a utilities trench plan—sometimes called a *trench composite plan*. These plans show where the utility trench will be located, cross sections of the various sections of the trench, and where transformers, pull boxes, and valves will be located. The trenches usually include all or some of the following: power, telephone, cable television, and gas. Street lighting is included in some jurisdictions.

Development plans

Some builders of single-family, detached homes use a drawing of the lots with a variety of information to assist in the construction of the homes or for acquiring permits. They may be required by the approving jurisdiction to verify that setback distances are met and individual lots drain adequately. These drawings may be called *development plans*. Do not confuse this with a planned development as described in Chap. 1. These plans are drawn to scale showing lot lines and streets. A footprint (outline of the building) is drawn on each lot. The development plan may be of a single lot or of several lots.

The building setback distances and drainage lines may be shown for laying out the foundation and acquiring permits. Other information that can be shown on each lot includes the model number of the house, the elevation type, colors, roofing, and siding type to be used. Any other information that will expedite the completion of the project can be included.

As-built plans

During the construction of a project, it is often necessary to make changes. If there is some unforeseen obstacle that prevents construction of the facilities as designed or there is an error in the design, a problem may exist that must be solved in the field. Any field changes should be shown on the construction plans. When the construction is complete and all changes have been marked on the construction plans, dated, and signed, that set of plans becomes known as *as-built* plans. Ideally, all changes made will be shown completely on as-built plans. In actual practice, this rarely happens. This is another reason why, even though as-built plans may be available, field surveys should be performed to verify locations before the engineer uses information from them to design other improvements.

Summary

Maps and plans are the work product of engineers. There are many choices as to how best to illustrate the work to be performed. Together the maps and plans show the approving agency, the public, and the contractors what the finished project will be and how its construction is to be accomplished.

Problems

1. What is the difference between a map and a plan?

2. What is the purpose of subdivision maps?

3. Name three ways plans are used.

4. What is the purpose of preliminary maps?

5. What is the purpose of tentative maps?

6. What are Conditions of Approval?

7. What is the difference between a parcel map and a subdivision map?

8. Name four elements of the cover sheet.

9. Define profile.

10. Define cross section.

11. What is the purpose of a master plan?

12. What is the purpose of a grading plan?

Further Reading

Consulting Engineers and Land Surveyors of California, Subdivision Map Act, Sacramento, California, 1989.

Preliminary Engineering

The first step of preliminary engineering is the site analysis. If a thorough site analysis was done, much of the preliminary engineering is also done. If not, the feasibility of the project should be investigated before more time and money are spent. Because of the pressure of time, engineers sometimes rush enthusiastically into the design of a new project without spending sufficient time evaluating and planning the work. Careful planning of the layout and presentation will result in a finished product in which the engineer can take pride. The preliminary engineering phase is the appropriate time for the planning to take place.

The purposes of preliminary engineering are to get approvals for the project from the appropriate jurisdictions and to evaluate the costs of construction. The cost estimate is very important and should be prepared by an experienced engineer. The expected cost of the project must be known before construction loans can be applied for and cash flow evaluated. Cost estimates may be requested several times during the development process. Each cost estimate will be more refined than the previous one.

People trained as planners often will prepare the street and lot layouts for the subdivision map or building site plans, but engineers should understand the criteria and be prepared if asked to design a building site or subdivision.

Design of alternative routes for linear projects should be prepared and evaluations made as to the various advantages and disadvantages of each. In addition to different locations, alternative routes can include routes with the same location but different cross sections and also a "do nothing" alternative. The size of the budget will limit the number of alternatives to be investigated. Alternative route plans are illustrated for display at public meetings so that the citizens affected can evaluate the impact on their lives and the environments.

Preliminary Design

The layout of the streets and lots for subdivisions, routes for linear projects, and site plans for commercial, industrial, or multifamily projects are affected by several factors. They are the locations of existing streets and utilities, the slope of the land, the presence of legal and environmental limitations, natural amenities, traffic flows, and criteria relative to lot sizes and setback distances.

The most economical use of a subdivision site will maximize the number of lots while minimizing the space occupied by streets. To accomplish this, several layouts should be designed. The minimum areas and frontages of lots may be dictated by zoning ordinances. Some developers have house plans designed with dimensions to fit frontage and side-yard setback requirements exactly, so that there is no unused frontage. Townhouses and patio homes are used to make the greatest use of street frontages and to reduce the relative cost of street and utility construction. Where there are existing or proposed easements, lots and streets should be designed so that easements fall within the street or at the edges of lots, so as not to encumber full use of the lots.

It is undesirable to use long, straight streets in subdivisions, as such streets encourage speeding. Also, curving streets are perceived as being more interesting. The cost of engineering, surveying, and constructing curved streets, however, is higher than for straight-line streets, so curves should be used judiciously. On hillside sites, the desirability or necessity of curvilinear streets is clear.

Street grades must be within an acceptable range for safety. Routes that curve around a hill rather than going directly to the top will not be as steep. Also, a street that follows the contours of the hill (Fig. 5.1), rather than going directly up the side (Fig. 5.2), will be less noticeable to those observing the hillside from below. Of course, a direct route will be less expensive.

When laying out the lots on hillside sites, allowance in the sizes of the lots must be made for slopes and drainage. This allowance cannot be estimated. Slope requirements and building setbacks must be determined. The cost of grading is a major expense. As much as possible, the natural contour of the land should be used, and split-level house plans should be encouraged on hillside locations. The appearance of the homes from the street level should please the eye, so cut or fill slopes of greater than 2.5 m (8 ft) should be avoided if they will be visible from the street.

Preliminary design of a highway route depends on a number of factors. In urban areas, the cost of land may be of primary concern. Demographics is also an important factor. Cutting through established neighborhoods may create serious social problems. Taking land occupied by low-income housing may require that housing be replaced. The more individuals whose lives will be affected, the higher will be the social costs. During the construction of the Aswan Dam in Africa, for example, whole villages were moved and resettled in exactly the same

Figure 5.1 Driveway following the contours of the hill.

spatial configuration at a new location as they had been at the original location.

The design engineer must be aware of the ownership and purpose of the buildings shown on topography maps. What appears as just a box on a topographic map may be a greenhouse made of wood framing and

Figure 5.2 Driveway going directly up the side of a hill.

plastic sheets with very little monetary value, or it could be a sophisticated research site valued at millions of dollars.

The extent of what property will be needed must include setback distances to other structures and right-of-way or property lines. The space needed must also include cut and fill slopes and possibly areas for reconstruction or realignment of drainage facilities.

In rural environments, earthwork may be the greatest factor. The steepness of the grades allowed for highways and railroads can determine the horizontal layout. The objective will be to align the highway to minimize and balance the earthwork. Channels for transportation of water must function by gravity flow as much as possible, to minimize the need for pumping systems. The use of bridges and other structures should also be minimized. Disturbance of natural resources such as wetlands and other unique natural habitats should be avoided. When wetlands and unique habitats are destroyed, they will have to be replaced—usually a very expensive process.

The need for field visits cannot be overemphasized. The more familiar the design engineer is with the project environment, the better the design will be.

Preliminary Engineering

When a preliminary design or alternative routes have been selected, the preliminary engineering can begin. If a tentative map is available with its conditions of approval, it should be used for preliminary engineering and cost estimates. The object of preliminary engineering is to do an initial layout of roadways, structures, storm drain facilities, sewers, and other utilities so that a cost estimate can be made. Do not let the preliminary nature of the task lead you to be careless in your analysis. The preliminary work could be accepted and refinement of calculations made without further analysis.

On small projects, it can be helpful to make the preliminary engineering layout on a print of the preliminary or tentative map. The plans or alternative layouts for linear projects should be as clear and complete as the information available allows.

Creating preliminary CADD drawings

Drawings showing preliminary layouts of proposed projects will be prepared for presentation and evaluation by governmental agencies. It is during this phase that preliminary layout of the engineering elements will first be put into the computer. The engineering information necessary at this stage will be graphic and time will not be spent showing exact calculations. The purpose of the drawings is to show that criteria for development is met and that the project can be made to work. The data shown at this phase is basic and minimal. The overall project is illustrated and is a master plan.

Before computers were widely used, the plans created during this phase were saved for reference for the final design but new drawings were made for construction plans. Now that the preliminary layouts are created in the computer, engineers are making use of them to create the final design. The problem with this approach is that whenever changes are made, the chance of errors is greatly increased. And the very nature of building the final design directly on the preliminary design is creating changes. This is further complicated by the fact that the others involved in the design will be making changes to their original concept as well.

Another problem that arises because of the use of CADD for preliminary drawings is that the way the drawings look on the screen is not the way they look in hard copy. This is because when we want to focus on a particular area, we simply zoom in on it and can see the desired area clearly. The fact that line work is in a variety of colors in computers but shades of gray in hard copy also is deceptive. The result is that what looks clear while working on the computer may look jumbled in hard copy unless there is consideration of these factors. One way to alleviate this problem is to plan to use larger-scale drawings.

Before any work is performed, engineers must take the time for careful consideration of how the final work product will be presented. If the work is being performed for a public works organization, it may have exacting standards of how each of the elements of the work is to be shown. That is, the organization will dictate the sheet sizes, borders, scales to be used, text size and type, as well as what interior sheets are required and what information must be shown on the cover sheet and interior sheets.

On privately funded projects, this is not usually the case. The team that will be creating the plans must take the time to decide a number of things. If the project is a complex site plan such as for a multifamily residential, commercial, or industrial development, computer information will be received from the architect and landscape architect and others. It is important that the information received does not complicate the civil engineering work unnecessarily.

The engineer must make it clear that though he or she can work with whatever is requested, some things can increase the engineering fee. Shared computer information should be kept as simple as possible. Drawings received from the architect should be limited to a footprint of the buildings. Inclusion of other information will have to be purged before the information is installed into the file of that project. Since several consultants will be receiving information from the architect, that same limited information will be true for others as well. Therefore, it makes more sense for the architect rather than the various consultants to remove extraneous information. All consultants should list what information is needed for their work and present it to the architect.

The site engineer needs only the outline of the buildings. The information should include locations of entrances and exits and any outside facilities that may affect grading. Information on floors that are not at ground level should not be included. The ground floor information should also be kept to a minimum. Locations of electrical and mechanical rooms and elevators may be useful but other information should be purged from the file. The fact that extraneous information takes up memory on computers may not be significant, but the time to scroll through architectural layers and then interpreting the layer names may be guesswork and a trial-and-error process. The architect should be able to purge unnecessary information in a short while. It could take many hours for the engineer.

What the architect gives to the other consultants will most likely also hold boundary information and possible street information received from the civil engineer. It should be on a separate layer and not bound together so that the boundary and street information can be turned off or purged when returned to the engineer. This way, if there are changes to the information originally provided by the engineer, the revised information can easily replace the superseded information.

The engineer provides the property boundary and street layout information to the architect. It is important for the engineer to provide information to the architect that will facilitate the architect's work as well. Any information such as survey points or topography not needed for the architect's work should be purged. Further, it is important that information given has been *twisted* or oriented by the engineer to the way that best fits how the architects want to present their plans. This way the engineer can be assured that coordinate and other mathematical information will be preserved. If the orientation is left to the architect, there is the risk that the civil information will be rotated in a way that loses the coordinates and that the architect's information and changes will have to be twisted every time information is sent back to the engineer.

The civil site plans and the architect's plans should be oriented the same way on the plan sheets unless there is good reason to do otherwise. Normally, civil site plans are oriented so that north is up or to the right on the sheets and CADD programs are constructed that way. If the architect or client chooses to orient the work so that north is down or to the left, it can be done, but the client needs to understand that it could slow the engineering and drafting.

Maps

Once a preliminary layout has been selected, a tentative map can be prepared and the approvals procedures begun. Where it will be necessary to install utility lines or ditches through property held by others, easements or rights-of-way must be purchased. To accomplish this, title reports should be ordered, legal descriptions prepared, and nego-tiations begun. This process should be started as soon as the need for the easement is realized. The smaller the investment made before all required property is secure, the less will be the risk involved and the better the client's bargaining position.

After the tentative map has been approved, there will be conditions of approval to be satisfied. The surveying calculations and drafting of the subdivision or parcel maps will be started. This procedure, as well as any legal descriptions, takes time and expertise, so there are costs associated with it. An estimate of how much it will cost should be made and the client informed.

On linear projects, drawings must be assembled to show the owner-ship of the lands affected. At this early stage, the drawings are usually made from information available from the assessors' parcel maps. It will be necessary to have all of the assessors' parcel maps converted to a common scale for assembling into a strip map (see Fig. 2.4).

There may be areas where the assessors' parcel maps cannot provide needed, current information as to such things as widths of streets. Where more information is needed, the subdivision or Record of Survey maps should be acquired. The information from these maps will be more reliable than information from the assessors' maps as to bearings and distances but not as to ownerships. The subdivision and Record of Survey maps will be needed later for use by the survey crews to deter-mine exact property ownership boundaries. The time and expense of acquiring all of the subdivision and Record of Survey maps at this early stage is not justified.

Clearing, grubbing, and demolition

On simple projects, clearing, grubbing, and demolition can be included on the grading plan. Where there are a number of structures to be removed or to be saved, a demolition plan is recommended. The cost of the demolition can be difficult to estimate because there may not be a history of the same unit costs from which to draw. Certainly, if the site is covered with orchard, clearing and grubbing will cost more than if the site is an open field. If there is paving to be removed from the site, demolition should be estimated on the basis of square meters. It may be appropriate to estimate clearing based on square meters.

Demolition of buildings should be estimated based on the number and sturdiness of the buildings to be demolished. A contractor expect-ing to be given the demolition contract may be willing to provide an estimate. Previously unimproved sites require removal of organic material from the soil. The depth will vary from site to site. The soils report should address this and recommend the depth of soil to be removed, but at this early stage the soils report is not likely to be avail-able. The depth used for other sites in the area can be used. Fortunately, the cost of clearing, grubbing, and demolition is usually a small portion of the total cost of construction.

Grading

Grading can be one of the most important aspects of a cost estimate. This is particularly true if the site will have to be built up and if earth must be brought in from kilometers away. It is possible that a contractor will pay to be allowed to dump excess material from another site. Most often, if material is needed, it will be expensive to locate, quarry, and transport to the site. In a metropolitan area, the cost of imported material can make a project infeasible. In most cases, the earthwork can be made to balance—the amount of excavation equals the amount needed for fill.

The cost of grading for streets and building pads is based on removing the plant material and compacting the soil to the specifications of the soils engineer. The cost of grading building pads will be charged at a different rate than the rate for grading streets and parking lots, so it should be listed separately on the cost estimate.

Streets and highways

The preliminary map for a subdivision and alternative maps for highway routes will provide the layout of the roadways. The standards of the jurisdiction for the geometric requirements indicate the widths of the traveled way and curb, gutter, and sidewalk requirements. The standards are based on the amount of traffic expected.

The structural section (see Fig. 7.19) is based on expected traffic and soil conditions. The requirements of the structural section will not be known in the early stages, but the preliminary estimate can be based on other roadways of similar use and width in the area. This information can be taken from adjacent projects. Be aware, however, that the structural section may not be indicative if the adjacent projects were built much earlier.

In subdivisions, there are often two or more types of streets within one project. A cul-de-sac may have a shallower section than a primary street. Because of this, each type of street will have to be estimated separately. The roadways can be shown as so many square meters of paving, such as 140 mm (5½ in) of AC (asphaltic concrete) on 250 mm (10 in) of AB (aggregate base) or 100 mm (4 in) of AC on 150 mm (6 in) of AB. On some projects the asphalt and aggregate may be shown in tonnes rather than square meters.

Storm drains

A rough estimate of the expected rainfall runoff must be made. The master plan of the storm drainage agency must be consulted. If the project is surrounded by previously developed sites, the runoff may be the result only of what falls directly on the project. Otherwise, an estimate of runoff from adjacent areas must be made. Determining the runoff area is described in Chap. 9, under the heading "Hydrology."

If an area master plan is not available, a USGS map can be used to delineate drainage basins. The project and those adjacent areas which will contribute to storm runoff over the project should be divided into drainage basins so that an estimate of the number and location of storm water inlets or culverts can be made and diversion ditches planned. The locations of manholes can be laid out. The sizes of pipes can be estimated based on the runoff and approximate slope of the ground. Need for bridges can be identified.

In some situations, it may be necessary to accommodate upstream runoff that will require the involvement of a flood control agency. That a site does not appear to have a water course does not mean that it is not subject to flooding. The Federal Flood Insurance maps must be consulted. Another cost factor is the federal Non-Point Source Pollution Controls requirements now being implemented to recharge the groundwater and protect our waterways from pollution. It is necessary to find out what will be required so that costs can be included.

If the project is linear, it may cross existing storm drain lines. This can be a major problem and a major cost element if the project will lie below existing ground, such as an urban freeway below grade or a construction of a major water supply or storm drain line. Existing storm drains will most likely be dependent on gravity flow. If there is a conflict in vertical location—that is, if the pipes are at the same elevation where they cross—the new one will have to be designed to go over or under the existing pipe. If that cannot be made to work because of other constraints, the existing line may have to be relocated vertically. Inverted siphons rarely work adequately and may not be acceptable to the storm drain agency. It may be necessary to install a pumping system or to transport the drainage long distances to accomplish the necessary clearance for crossings. These are major costs that can be critical and must not be missed at this stage of planning.

Sanitary sewers

For new subdivisions, the initial layout of sanitary sewers is simpler than that of storm sewers, since every building must be served. If the mains will serve only the buildings on the project and it is a residential site, minimum-size laterals may be adequate throughout. The responsible agency dictates the flows to use for each household, a *peak flow* factor, and the infiltration rate. Manholes and flushing inlets should be included in the layout and cost estimate.

It is important to contact the agency responsible for the area's sanitary sewer master plan to determine that the outfall you plan to use has capacity available for your project. Also, it may be necessary to install a sewer main with a capacity far exceeding the need of your subdivision to accommodate future development. If excess capacity is required, there should be cost sharing by the agency.

Linear projects crossing existing sanitary sewer lines must consider the same concerns as previously described in the section on storm drains. These are critical factors in determining costs.

Water lines

If the water supply company will be installing the water systems, the cost of providing water services should be based on their estimate. If the responsibility for the design of the water supply is with the engineering consultant, an estimate must be made. A layout of the lines, remembering loops, should be made. Some water companies have significant charges if a line is to have a wet tap. Mains, laterals, water meters, fire hydrants, air-relief valves, thrust blocks, and anchors must be included in the estimate.

If you are designing a linear facility which crosses small water supply lines, crossings are rarely a problem. Water supply lines are forced systems—that is, the water is transported under pressure. For this reason, the water lines are not dependent on gravity flow and will function even when designed to flow uphill. Clearances are not as critical, and the water line can be relocated over or under the new facility. There will be costs involved which must be accounted for, but it is unlikely that these costs will be critical. However, if the water supply line is a major trunk line, the cost of relocation, whether horizontally or vertically, will be a very high-cost item. The agency that operates the trunk line must be contacted to determine what its position will be and what costs to expect.

Other utilities

Other utilities, such as power, gas, and communications lines, may have to be extended to your site. There may be a requirement that overhead power and communication lines be relocated underground. This can be very expensive. There may also be a need for transformers and substations, which may increase costs and require easements. Some utility companies pay for the installation of the utilities; some look to the developer to pay costs.

Crossing these existing utilities by linear projects is usually not a problem for the design engineer, as the utilities can be placed almost anywhere. However, there may be costs associated with relocations.

Traffic markings and safety features

The cost of installing traffic signals, signing, lighting, striping, and other traffic markings and guard rails should be included. An estimate of the number needed should be based on the type of features and placement required by the jurisdiction.

Cost Estimating

Estimates of the costs of construction have to be calculated throughout the project so that the client can have some control over cash flow and be prepared if expected costs change. When the initial layout of the streets and utilities is complete, the preliminary estimate can be prepared. It is often helpful to list the quantities on the same sheets used for the preliminary layout of utilities. This way, if the final cost of construction is different from the preliminary estimate, a comparison of designs can be made and the reason for the difference determined. Most engineering firms have a list of items that might be necessary to include (Fig. 5.3). By using the list, items are less likely to be missed. Of course, there may be items to include which are not on the standard form.

Count the various items needed, measure the lengths of conduits, and calculate the square meters of paving and other quantities. Accuracy is important, but the plans will not be complete enough to make precise measurements. If a length of pipe is between 180 and 190 m, use 190 m. If there is a minimal amount of grading of the site needed and the quantity is to be in cubic meters, assume earthwork as 0.6 m (2 ft) of depth over the site. Be conservative. Clients get loans based on your estimates; if they are too low, the client may be in financial trouble. On the other hand, if your estimate is high and you cannot justify it, clients will take their business elsewhere.

Compare the estimate with similar projects. If the costs vary significantly, be sure the differences can be justified. When the initial layout of the streets and utilities is complete and the quantities have been estimated, the preliminary estimate can be prepared. The estimate should be identified at the top, for example, "Engineers Preliminary Cost Estimate for Carlsen Property, based on a 1:1000-scale preliminary site plan dated June 8, 1998." It is essential to show the date, as preliminary plans are subject to change and you must be sure the plan being used is the correct one. Also, this way, if the final plan is significantly different and the costs differ from the estimate, the reason for the difference can be understood.

The format used will vary among engineers, clients, jobs, and times. Each estimate should be tailored to the job. There may be reasons to segregate the project into parts for the estimate. If the project is large, it may be constructed in phases. It may be desirable to separate the on-site and off-site work. Various conditions, such as an adjacent property being developed, may change with time, so required work and costs will vary from time to time. This should be explained in the notes section and comparison of costs with time listed. (See Fig. 5.4.)

Date: _____

Job No. _____

PRELIMINARY COST ESTIMATE

Quantity	Units	Description	Unit Price	Subtotal	Total
Planning					
	L.S.	Annexation			
	L.S.	Zone Change			
	L.S.	Preliminary Planning			
	L.S.	Tentative Map			
		Other			
Subtotal					
Surveying					
	L.S.	Boundary Survey			
	L.S.	Control Survey			
	L.S.	Aerial Survey			
	L.S.	ALTA			
	L.S.	Final Map			
	L.S.	Easement Descriptions			
	L.S.	Construction Surveys			
		Other			
Subtotal					
Grading					
	L.S.	Demolition			
	L.S.	Clearing and Grubbing			
	m^3	Removal of Hazardous Materials			
	m^3	Rough Grading			
	m^2	Street grading			
	m^2	Lots			
	m^2	Ditches			
		Other			
Subtotal					
Paving					
	m^3	asphaltic concrete			
	m^3	aggregate base CL 3			
	m^3	cement treated base			
	m^2	geotextile			
	m	curb and gutter			
	m	vertical curb			
	m	AC berm			
	m^2	PCC sidewalk			
	each	driveway approach			
	each	handicap ramps			
	LS	signing and stripping			
	LS.	traffic signals			
		other			
Subtotal					
TOTALS PG 1					

Figure 5.3 Preliminary cost estimate form.

Date: _____
Job No. _____

PRELIMINARY COST ESTIMATE

Quantity	Units	Description	Unit Price	Subtotal	Total
Sanitary Sewers					
	each	100 mm (4 in) laterals			
	each	200 mm (8 in) laterals			
	m	200 mm (8 in) VCP			
	m	200 mm (8 in) PVC			
	m	250 mm (10 in) PVC			
	each	standard manhole			
	each	manhole w/ outside drop			
	each	cleanouts			
		other			
	Subtotal				
Storm Drain					
	each	flat grate inlets			
	each	catch basins			
	each	standard manhole			
	each	outfall structure			
	each	concrete lined ditch			
	m	250 mm (10 in) PVC			
	m	900 mm (36 in) RCP CL V			
	m	100 mm (4 in) perforated pipe			
	each	clean out			
		other			
Subtotal					
Water					
	m	250 mm (10 in) PVC water main			
	each	25 mm (1in) PVC lateral			
	each	38 mm (1.5 in) PVC lateral			
	each	25 mm (1in) water meter			
	each	38 mm(1.5 in) water meter			
	each	wet tap			
	each	fire hydrants			
	LS	thrust blocks & fasteners			
		other			
Subtotal					
Miscellaneous					
		Sanitary Sewer Connection Fee			
		Storm Drain Connection Fee			
		Joint Utility Trench			
		SWPPP			
		City Inspection Fees			
		Engineering Fee			
		Contingencies			
		other			
Subtotal					
TOTAL PG 2					
TOTAL					

Figure 5.3 *(Continued)*

Date: June 8, 1998
Job No. 17581C

PRELIMINARY COST ESTIMATE
Johnson Property (APN 640-45-78&79)

Quantity	Units	Description	Unit Price	Subtotal	Total
Planning					
	L.S.	Preliminary Planning		$ 6,000	
	L.S.	Tentative Map		$ 8,000	
Subtotal					$ 14,000
Surveying					
	L.S.	Boundary Survey		$ 21,230	
	L.S.	Control Survey		$ 6,500	
	L.S.	Aerial Survey		$ 19,650	
	L.S.	Final Map		$ 18,900	
	L.S.	Construction Surveys		$ 23,560	
Subtotal					$ 89,840
Grading					
	L.S.	Clearing and Grubbing		$ 6,400	
52,300	m³	Rough Grading	$ 2.10	$ 109,830	
3,660	m²	Street grading	$ 1.60	$ 5,856	
21	each	Lots	$ 600.00	$ 12,600	
510	m²	Ditches	$ 1.50	$ 765	
Subtotal					$ 135,451
Paving					
3,660	m³	60 mm (2.5 in)asphalt concrete	$ 3.50	$ 12,810	
3,660	m³	250 mm (10 in) CL 3 A.B.	$ 3.50	$ 12,810	
380	m	curb and gutter	$ 2.10	$ 798	
380	m²	PCC sidewalk	$ 12.00	$ 4,560	
21	each	driveway approach	$ 180.00	$ 3,780	
4	each	handicap ramps	$ 220.00	$ 880	
	LS	signing and stripping		$ 800	
Subtotal					$ 36,438
Sanitary Sewers					
21	each	100 mm (4 in) laterals	$ 60.00	$ 1,260	
180	m	200 mm (8 in) PVC	$ 5.00	$ 900	
1	each	standard manhole	$ 2,000.00	$ 2,000	
1	each	manhole w/ outside drop	$ 2,600.00	$ 2,600	
Subtotal					$ 6,760
TOTALS PG 1					$ 282,489

Figure 5.4 Preliminary cost estimate form filled out.

Factors affecting costs

A unit cost will be allocated to each item on the cost estimate listing. These unit costs will vary from job to job. Costs are affected by the following:

1. *The size of the job.*   The unit cost of material and installation for a large job is less than for a small job. The overhead costs and the cost of transporting equipment and material to a site will often be nearly the same for 2 units as for 22 units.

Date: June 8, 1998
Job No. 17581C

PRELIMINARY COST ESTIMATE
Johnson Property (APN 640-45-78&79)

Quantity	Units	Description	Unit Price	Subtotal	Total
Storm Drain					
2	each	flat grate inlets	$ 550.00	$ 1,100	
2	each	catch basins	$ 630.00	$ 1,260	
2	each	standard manhole	$ 2,100.00	$ 4,200	
150	m²	concrete lined ditch	$ 9.00	$ 1,350	
170	m	300 mm (12 in) RCP CL V	$ 12.00	$ 2,040	
150	m	100 mm (4 in) perforated pipe	$ 4.00	$ 600	
1	each	clean out	$ 100.00	$ 100	
Subtotal					$ 10,650
Water					
170	m	100 mm (4 in) PVC water main	$ 6.00	$ 1,020	
21	each	25 mm (1in) PVC lateral	$ 30.00	$ 630	
2	each	38 mm (1.5 in) PVC lateral	$ 34.00	$ 68	
21	each	25 mm (1in) water meter	$ 250.00	$ 5,250	
1	each	38 mm(1.5 in) water meter	$ 280.00	$ 280	
1	each	wet tap	$ 400.00	$ 400	
2	each	fire hydrants	$ 1,500.00	$ 3,000	
	LS	thrust blocks & fasteners		$ 1,500	
1	each	blow-off valve	$ 210.00	$ 210	
Subtotal					$ 12,358
TOTAL PG 2					$ 23,008
TOTAL PG1					$ 282,489
TOTAL CAPITOL IMPROVEMENTS page 1 & 2					$ 305,497

$305,497/21= $ 14,547 Per Lot

Quantity	Units	Description	Unit Price	Subtotal	Total
Miscellaneous					
		Sanitary Sewer Connection Fee		$ 10,500	
		Storm Drain Connection Fee		$ 5,000	
		Joint Utility Trench		$ 50,000	
		SWPPP		$ 5,000	
		City Inspection Fees		$ 15,000	
		Engineering Fee		$ 30,000	
		Contingencies		$ 39,715	
		School Fees		$ 42,000	
Subtotal					$ 197,215
TOTAL					$ 502,712

$502,712/21= $ 23,939 Per Lot

NOTES:

1. This estimate is prepared as a guide only, is based upon incomplete information and is subject to possible change. Carter Engineers, Inc. makes no warranty expressed or implied that actual costs will not vary from the amounts indicated and assumes no liability for such variances.
2. The estimate does not include:
  a. Soils Engineering Costs
  b. Landscaping
  c. Fencing
  d. Building related costs and fees
  e. Reimbusible agreements or refundable deposits
3. Costs are based on estimated current prices with out provisions for inflation.

Figure 5.4 (*Continued*)

2. *The location of the site.*   Again, transportation costs depend on distance and accessibility of the site. This is particularly true in the case of imported earth.

3. *The client.*   Some public agencies are notorious for elaborate specifications and fastidious inspectors. Contractors know that these factors will be costly in time and money.

4. *The engineer.*   If incomplete or incorrect plans have cost a contractor time and money on a job previously designed by a particular engineer, the contractor may be reluctant to work with that engineer again or may pad the costs with a large contingency factor.

5. *The season.*   Longer days and dry weather allow work to be accomplished more quickly, thus at less cost.

6. *The economy.*   If the economy is booming and contractors have all the work they can handle, the estimate may be high because there is not much incentive to do more work. If economic times are poor or a contractor is trying to establish a new business, the contractor may estimate lower to get the job.

7. *Financial factors.*   Cash flow is often an important factor. If payments are to be based on work completed, the unit costs of work done in the early stages of construction may be high, and those of work done in the late stages may be relatively low. This gives the contractor working capital at the beginning of the project.

8. *Cost of property.*   If the project is to develop property owned by your client, this is not a factor except if easements will be required across adjacent property. However, the cost of a right-of-way in a metropolitan area for a new freeway can be the single greatest cost and can be the most important determining factor.

The value of property is affected by its zoning and use. Residentially zoned property will probably be more valuable than property zoned agricultural; commercial and industrial property will be more valuable than residential property. Whether the property is vacant or developed also makes a significant difference in its value. A property appraisal company should be brought in to determine the value of the various types of property.

If it will be necessary to relocate homes and businesses, a company experienced in locating comparable property and familiar with the costs associated with relocations should be assigned the task of estimating costs. It should provide a formal report of the costs that can be anticipated.

Keep these influences in mind when selecting unit prices to use for the estimate. The values you select to use for unit prices when making estimates should be based on recent experience. For reference, keep a history of estimates you have made. Whenever possible, get the prices used by the contractors who bid on the projects you work on. This information may be the best source for estimating unit costs.

Some developers and public agencies use computerized spreadsheets which list the unit price estimated by contractors for each item to be used. Where quantities are similar, take unit prices from the spreadsheet for another job. When the job includes an unusual structure, it may be necessary to ask a contractor involved in that type of work for an estimate. One other source to use in determining the value of contract items is one of the books published each year which lists unit prices of construction material. Be sure to compare the unit prices suggested with the quantities being used and take into account an inflation factor if the value of money may have changed since the publication was released.

Miscellaneous costs and fees

There are always some costs and fees that are not for material and installation and do not fit into the other listed categories. There are fees to be paid for connection to storm and sanitary sewers and water systems. There may be permits for encroachment into streams or highway rights-of-way. The cost of putting power and telephone lines underground may be listed here. Some areas may have fees to help offset increased costs of fire and police protection and for schools and parks. There may be a variety of bonds required, which will add costs to the project. These are usually relative to the cost of improvements. Be sure to list them and include information as to what authority is imposing the cost.

Fee estimates

The cost of engineering and other professional services is called a *fee*. There will be a fee from the design consultants and from the reviewing jurisdictions. These fees are usually based on, or at least relative to, the cost of capital improvements.

Estimating the fee is one of the most critical tasks the engineer faces. It is also one of the most difficult. If the fee is overestimated, the client may seek another engineer. If the fee is underestimated, there may be no profit, or worse, a loss. If that happens often, the engineer will be out of business.

As with any professional service, for the client to make a choice based on fee alone is terribly risky. The engineer with the higher fee may have a better understanding of what is required and may design a project for which the cost of construction is much less. The choice of engineer should be based on qualifications, experience, and reputation, not the size of the fee.

The only way to be truly successful at fee estimating is to have extensive current experience. However, there are ways in which engineers can narrow the risks. By estimating the fee using three different methods and comparing the results, engineers can improve their confidence in the estimate.

The first step is to determine as nearly as possible the number of sheets that will be needed to illustrate the design. When the scale has been decided upon, the number of sheets that will be required to cover a particular area can be estimated. Be generous in your estimate. There is little advantage in using fewer sheets, and clarity may require more. There are usually a number of configurations that can be used. Be generous with space. It is false economy to try to get as much on one sheet as possible. Once the number of sheets needed has been determined, decide which sets of plans you will need to include. Use the list on pages 80 and 81 to help you decide. Some public works agencies require that quantities be shown as part of the set. Determine if that is the case and, if so, how many sheets will be required for quantities.

When you have a total number of sheets, multiply the total number of sheets times a cost per sheet for a similar project. There is no rule of thumb for determining the cost per sheet. The value will vary significantly from an uncomplicated private project to a complicated public works project and depending on whether specifications are to be included. This is another case where there is no substitute for experience.

Having completed that task, you will have a good understanding of the complexity of your job. The second approach is to estimate the number of hours that the engineers and other personnel will have to spend to produce the plans and specifications needed. Remember to include time for meetings with the client and other professionals, time for public hearings, and travel time.

If you have made a preliminary estimate of the cost of construction, you can use it as a guideline to how much your fee should be. The fee will generally fall somewhere between 4.5 and 12 percent of the net cost of construction—that is, the actual cost of construction not including engineering fees, permits, and bonds.

Having estimated the fee these three ways, compare the results. The results should be close. If they are not, you will want to reevaluate to be sure that you haven't left something out. Once the numbers are close enough to allow sufficient confidence, you may want to pick a number somewhere between the high and the low. If, however, the client is known as a tough negotiator, you may want to pick the highest number you can justify, to allow room for some negotiation. On the other hand, if this is a new client whose business you are anxious to have, or a client that you have worked for before and know to be good to work with, you may want to use a lower number.

Contingencies

Contingencies include unforeseen expenses and are usually set between 5 and 30 percent of the total cost of improvements, depending on the difficulty of the project and the likelihood of unanticipated costs.

Notes

The notes to the cost estimate are particularly important. Any assumptions used in making the estimate should be clearly explained. State whether the costs used are at current value or include an inflation factor. Most engineers include a disclaimer that the estimate is only a guide, and they assume no responsibility for variances from actual costs.

State possible costs that are not included, such as sound walls, environmental impact reports, and soils reports. Where the facilities required are oversized, the jurisdiction may give partial reimbursement to the developer for improvements larger than what is needed to serve the project. When this is the case, the anticipated reimbursement should be described. Costs that constitute refundable deposits, such as are sometimes required for water meters until the building is occupied, should be listed. Try to anticipate questions the developer may have, and provide the answers in the notes. Where there are unusual costs, include the name of the official stating the requirement.

Summary

The purpose of preliminary engineering is to acquire approvals of controlling jurisdiction, to determine what problems will affect the project, and what the cost will be to build it. The plans for the project will ultimately be a refinement of the preliminary engineering. At this point, the first cost estimates are made. These costs may be estimated again throughout the design process, but thoroughness is necessary from the beginning, as the success of a project is dependent on the cost estimates made.

Problems

1.   What are the purposes of preliminary engineering?

2.   Why is it important for architects and civil site designers to specify what should be included on exchanges of computer files?

3.   What is usually the highest cost factor in designing a linear project in an urban location?

4.   What is usually the highest cost factor in designing a highway in a rural environment?

5.   Why is it necessary for the design engineer to visit the site before making a preliminary cost estimate?

6.   If critical property ownership information is not clear on assessors' parcel maps, what resources can be used?

7.   How can you tell if the site is subject to flooding?

8. When a linear project crosses existing storm and sanitary lines, how are the crossings handled?

9. When a linear project conflicts with existing water supply lines, how are the crossings handled?

10. How do you determine the cost of property to be purchased for a linear project?

11. How do you determine unit prices for the estimate?

12. List four possible costs that would be included in a miscellaneous costs section.

13. What are the three approaches to estimating a fee?

Further Reading

American Society of Civil Engineers, *Consulting Engineering: A Guide for the Engagement of Engineering Services,* New York, 1988.

California Department of Transportation, *Contract Cost Data: A Summary of Cost by Item for Highway Contracts,* Publication Distribution Unit, Sacramento, California, 1992.

California Division of Mines and Geology, *Active Fault Mapping and Evaluation Program,* Sacramento, California, 1976.

Greene, L., Senate Bill No. 2034, Chapter 10 16, State of California, 1988.

6

Earthwork

Before concrete can be poured and structures built, the land must be prepared to provide a strong base. A civil engineer specializing in soils should be assigned to determine the characteristics of the soil and what construction methods should be used to provide the base for the structures. If the site is in a mountainous area or an area subject to earthquakes, a geologist or geologic engineer should also be contracted to evaluate risks and make recommendations for protection against landslides and earthquakes.

Surface and subsurface water can damage the base and cause failures as well. Therefore, each project must be sculpted and compacted to direct drainage away from buildings and other structures. The activities necessary to accomplish this are called *earthwork*.

The Geologic Report

Geologic reports should be prepared for projects in mountainous or earthquake-prone areas. The geologist will research the geologic history of the site, study aerial photographs, perform soundings to determine subsurface densities, and dig trenches across suspected earthquake faults and ancient landslides. Earth cores will also be extracted and studied. With this information, recommendations can be made as to areas where structures are at risk.

Landslides

Peoples' lives and property can be destroyed very quickly by landslides; therefore, areas of existing or potential landslides should be identified. Once a previous or potential landslide area is identified, recommendations can be made to avoid the risky areas. In some cases, areas of potential landslides or of soil creep can be used if certain precautions are taken or the structures are designed to accommodate the problems.

Groundwater

The geologic report may also identify groundwater conditions. If the water table is near the surface, it can be destructive to pavements or create problems to other structures. The geologist can make recommendations as to the scope of the problem and make suggestions for removing the water so that it will not adversely affect the structures.

A common method for removing groundwater is the use of french drains (Fig. 6.1). They are ditches in which permeable material is placed. The permeable material is wrapped in a geotextile to keep silt out. Groundwater moves into the ditch because the water can move through the permeable material more easily than the surrounding soil. A perforated pipe is placed in the ditch with the holes down and hydrostatic pressure forces the water into the pipe where it is then carried to a drainage system.

If a well is needed as a water supply source, the geologist can recommend locations for drilling based on experience and studies of the site. Sometimes there is a perched water table (Fig. 6.2). In this condition, groundwater is trapped by an impervious soil layer. If a well driller finds this layer and stops, the water may provide a limited supply of poor-quality water. The geologist should be able to identify the condition and may recommend drilling deeper or elsewhere. A deeper layer of groundwater will most likely have a better potential for a good supply of high-quality water.

Earthquakes

Earthquakes can be a threat to life and can damage or destroy structures. There are two primary ways that earthquakes cause damage. One is the lateral forces created when the earth moves, which is a structural engineering problem. The other is that some soils liquefy during the ground shaking. These soils and their depths must be identified so that foundations can be designed to overcome the liquefaction.

In most cases, structures can be designed to withstand the forces of earthquakes. The advances made in understanding the forces and development of methods to protect against lateral forces on structures

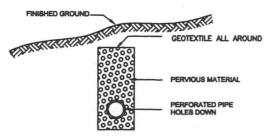

Figure 6.1 French drain.

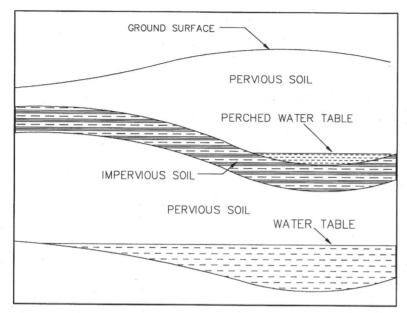

GROUND SURFACE

PERVIOUS SOIL

PERCHED WATER TABLE

IMPERVIOUS SOIL

PERVIOUS SOIL

WATER TABLE

Figure 6.2   Perched water table.

caused by earthquakes is testament to the work of engineers and geologists studying earthquakes. Considering the density of population in the area, the loss of life was very low (49 people) from the Loma Prieta earthquake in 1989 in California. Earthquakes of the same magnitude in other parts of the world kill thousands of people.

The Soils Report

An investigation of the soils should be made for every site. A report of the investigation should be made by a qualified civil engineer specializing in soils science. The soils engineer will visit the site, take soils samples, and make borings at various locations. The cores resulting from the borings show the underlying strata. A three-dimensional view of the layers of earth and rock can be projected from the cores. Though subsurface conditions cannot be described with absolute certainty, the unknowns are reduced and much useful information is provided.

The different types of soil and rock on the site are identified. A series of tests are performed on the soils to determine their strength, plasticity, potential for liquefaction, and permeability. The depth of groundwater is also provided. The level of groundwater varies with the time of year and the character of the previous rainy seasons. If the seasons have not been typical or there is historical evidence that groundwater is a problem, further investigation is indicated.

The information provided will be useful to the architect and structural engineer in designing the structures, to the site engineer in designing paved surfaces and slopes, and to the contractor charged with grading the site. If subsurface conditions change abruptly under a proposed structure location, it may be necessary to excavate existing earth to provide a consistent earth foundation beneath that structure or to design a different foundation for different parts of the structure. The site engineer should read the soils report in its entirety before beginning the grading plan.

Slopes

The report should describe maximum allowable slopes. The allowable slope is based on the angle of repose for the soil. Using a steeper slope could result in slope failure or landslide. The *slope* is described as the unit vertical distance necessary for each unit of horizontal distance. For a slope of 1:2 there is 1 unit of vertical difference for every 2 units of horizontal difference (Fig. 6.3). These slopes will be used between adjacent building pads, roadbeds, or other structures of different elevations.

Where there will be high cut or fill slopes, benches (Fig. 6.4) are usually required in the slope. The benches will stop falling rocks and earth and will be used to intercept and redirect overland drainage. Benches are also required in existing ground that has slope where there will be embankment placed over it. The natural slope is scraped clean of any organic material then cut into benches. The vertical distances between benches and the width of the benches is determined by the character-

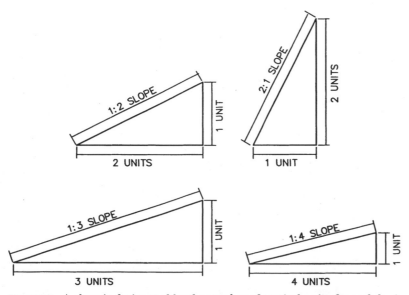

Figure 6.3 A slope is designated by the number of vertical units for each horizontal unit. (English system uses horizontal units for each vertical unit.)

Figure 6.4 Cut slope benches.

istics of the soil, widths needed to operate equipment, and what the finished slope will be. Benches so employed usually also have a key in the bench to connect the soil masses. (See Fig. 6.5.)

Compaction

The soil compressibility should be included in the soils report and is important to the site engineer. Typically, to provide an engineered base for structures, the soil must be 95 percent compacted. The natural earth in place may not be compacted to 95 percent, in which case more earth will be required to fill the same space after compaction. A clear demonstration of this can be seen by filling a cup loosely with sand and clearing off the excess sand level with the top of the cup. If you then tap the cup several times, the sand will compact, and the cup will no longer be full. The same is true for earthwork. The soils report should describe the optimum moisture needed and construction methods to be used to accomplish the recommended compaction.

All sites require some excavation and some filling. If the earthwork is measured in cubic meters for design and estimation purposes, more than a cubic meter of excavation will be required for each cubic meter to be filled. The percentage difference, expressed as a portion of 1, is called the *compaction* or *shrinkage factor*.

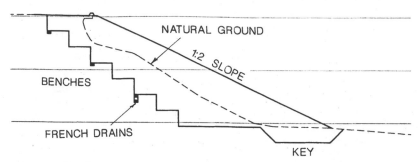

Figure 6.5   Benches are cut into the existing ground to provide a stable base for fill material.

It is desirable to have the excavation and fill on a site balance. The earthwork on a site is said to balance when no import or export of material is required. To accomplish a balance, a volume of earth to allow for shrinkage must be included in the calculations.

Paving

Structural sections for streets and parking lots should also be recommended in the soils report. The structural section for a roadway will be based on the volume and weight characteristics of the traffic expected. This is called the *traffic index* (TI). There will be a range of possibilities. Thicker asphaltic concrete (AC) requires less thickness of aggregate base (AB). Cement-treated base (CTB) may be used in place of, or in conjunction with, aggregate base. Where practical, list the alternatives on the plans or in the specifications. Where the native soils have poor structural qualities or are expansive, the soils report may recommend importation of soils better suited to providing a subbase for structures.

Where groundwater may be a problem, a pervious layer wrapped in a geotextile may be recommended to remove the moisture and protect the integrity of the base.

Sources of Earthwork

There are several types of earthwork activities that must be performed to construct the base for any project. They are removal of unsuitable material, excavation and embankment, miscellaneous earthwork, and shrinkage.

Unsuitable material

Most sites have at least some material that is unsuitable for engineered fill. It may be debris dumped on the site from other sites, or it may be the naturally occurring topsoil containing organic matter. Organic material is unsuitable for compacted subbase and must be scraped from the site before construction of the base is begun. The depth of material to be removed should be recommended in the soils report. Even 0.08 m (0.25 ft) of material to be removed over the entire site is significant and should be listed on the earthwork tabulation. The organic material and soil is usually stockpiled on the site for use in landscaping.

Constructing the base

Earthwork is a major factor in the cost of most projects. One of the objectives in designing any project is to minimize the earthwork. On projects that will be constructed on relatively level terrain, this objective is usually easily obtainable. In mountainous areas, however, other design criteria may dominate. The design engineer will have to work to minimize the volume of earthwork while considering other criteria.

Building sites should be designed to be level, with no more variation in slope than is sufficient to drain the property and provide landscaping. Highways and railways must not exceed certain grades if they are to provide safe, efficient traveled ways. Channels require a gentle slope.

In mountainous locations, buildings can be built with split-level foundations and sloping or stepped-down parking lots to minimize grading. To accommodate linear projects, mountains of earth must be moved. It is the cutting down of hills and filling in of valleys that generates the greatest amount of work and costs. The earth excavated in cutting down a hill is transported to a valley for placement.

When the unsuitable material has been removed, the major earthwork can begin. On linear projects, heavy equipment begins the excavation by cutting a wide swath where the finished top of slope will be, then continues cutting down layer by layer, with each layer narrower than the last, to provide the required side slopes. The earth is then transported to a location where it can be used for fill. The side slopes are dressed to provide a smooth, even slope.

When fill dirt is to be placed to bring up the elevation, whether for a building pad or a highway, the area to be filled requires preparation to ensure stable placement and a firm connection between the existing and new ground. Organic and other unsuitable material must first be removed. Any holes that resulted from removal of trees, structures, wells, cesspools, or septic systems must be filled and compacted to 90 to 95 percent. Level steps or benches (Fig. 6.5) must be cut into the receiving hillside. The fill is placed in courses of a depth recommended by the soils engineer. It is watered to provide the optimum moisture for compaction as recommended in the soils report. Each course is tested to determine if the compaction is sufficient before the next course is placed. This method is repeated until the desired elevation is reached.

Miscellaneous earthwork

Where buildings are to be built on the property line, the pad may have to be overcut or overfilled along side yards to provide space for workers and equipment (Fig. 6.6). On residential sites, the entire lot may be made

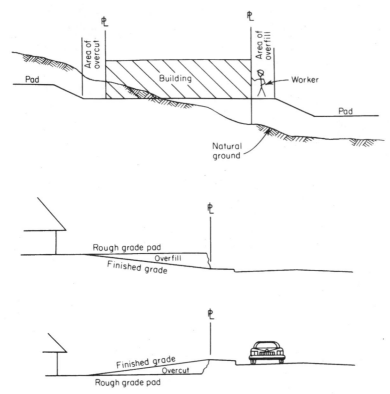

Figure 6.6   Areas of overcut and overfill.

level at the rough-grading stage. If the grading is handled this way, and finished grading is complete, a wedge-shaped section is cut away from the pads above the street and filled in on pads below the street. These wedge-shaped areas are also referred to as areas of overcut or overfill and must be accounted for when listing the quantities of earth to be moved.

The earth taken from trenches for utilities is called *trench spoil*. The trench is often then filled in with imported material that provides special characteristics, so the trench spoil is not replaced. This spoil is usually not significant in calculating earthwork for residential and commercial or industrial sites, but be aware that it is a significant factor on pipeline projects and may be a factor where unusually large trenches are required.

Shrinkage

As previously described, the volume of earth excavated from one area may not provide the same volume of earth for fill or embankment. The difference is called *shrinkage*. When the shrinkage factor is applied, the actual compacted earth required can be determined. That volume of earth is a source of earthwork. The method to determine the earth needed to compensate for shrinkage is described under "Determining Earthwork Quantities."

Designing Grading Plans

The purpose of the grading plan is for the engineer to design a firm foundation for structures and protect people and the structures from drainage problems. The plan is used to provide an economical design for construction and to communicate to the surveyors and contractors what is to be built.

Survey crews will use the plan to place 2×2 wooden hubs (guineas) and stakes. The elevation of the top of the hub is determined by the survey crew. Each survey stake is labeled with a brief description of what the hub marks and the amount to cut or fill from the top of the hub. "ER TC, C 380 mm" (1.26 ft) indicates that the hub is located at the end of the curb return (ER), and that the earth must be cut away 380 mm (1.26 ft) below the top of the hub to the top of curb (TC) elevation. The back of the stake will carry information describing the horizontal location, such as "1.7 m Rt CL STA 20 + 00" (56 ft Rt CL STA 20 + 00).

A construction crew member called a guinea hopper will read and call out to the equipment operator whether to cut away or fill in earth, and the depth to cut or fill. The engineer must be sure that there is enough information on the plan so that the field crew will have no problem doing its job. It is much more efficient for the engineer to calculate information while creating the plan than for the survey crew chief to calculate it while the other members of the crew wait.

Elevations

Of prime importance in understanding the various elements of the grading plans as well as the other aspects of design is the concept of elevations. When the term *elevation* is used, it may refer to an actual elevation (vertical distance in meters or feet above mean sea level), or it may refer to a vertical distance above an assumed elevation. Though the dimension of the elevation is in meters or feet, it is customary to show elevations without a dimension.

All plans using elevations should have a benchmark (BM). The benchmark is a vertical reference point. The benchmark may be a brass disk set in concrete by the U.S. Geological Survey (USGS) or some other agency, and tied to mean sea level, but it can be anything that has a permanent elevation that can be referenced. Some jurisdictions require that all plans be referenced to their standard benchmarks or USGS benchmarks. At this writing, USGS maps and benchmarks are in English units (feet) except for some of the 1:100 000 maps produced in 1991 and 1992. Whether the elevations are in feet or meters will be clear from information provided on the map.

On projects where there is no existing benchmark in the vicinity, the surveyor may establish a benchmark using some permanent feature such as a top of curb or manhole cover and give it an arbitrary elevation high enough so that no point related to the project will have a negative elevation. This point then has an assumed elevation and elevations are given to elements needed to design and build the plan in reference to the benchmark. What is important is that all the vertical relationships among the design elements is established. There are areas where the land is below sea level and will have negative elevations, but when an assumed elevation is to be used for the benchmark, negative elevations should be avoided.

Care should be taken when using elevations from existing plans. The benchmarks used to design different projects are often taken from different sources, so the relation between elevations on the projects will not be true. The elevation for a physical object taken from one benchmark may be different from an elevation for the same object taken from another benchmark, unless the two benchmarks refer to a common benchmark. Even then, there may be some differences due to the degree of precision or errors. Where two or more sets of existing plans are to be tied together, it may be necessary to establish a benchmark equation. An example is:

Rim elevation for sanitary manhole on Main Street at Spring Street =

139.68 from Tract 5555 and = 140.03 from Tract 5560

In this case, if elevations for Tract 5560 are to be used on the new project, but ties must be made to objects in Tract 5555, 0.35 (140.03 − 139.68) must be added to all elevations taken from Tract 5555.

Before design is begun on the grading plan, elevations should be shown wherever they must be considered in the design. This includes elevations for the following existing and proposed items:

1. Natural ground

2. Ditch flow lines within project boundaries and in some cases outside a sufficient distance to show the limits of the drainage basin (Chap. 9) contributing flow to the project

3. Tops of curbs at

   a. Property lines

   b. Beginnings and ends of horizontal curves

   c. Beginnings, ends, and high or low points in vertical curves

   d. High and low points in street center line profiles

   e. Points beyond the property line as necessary to show the grade of the street so that smooth transitions can be made

4. Existing streets being met at connections and as necessary to show the grade of the street so that smooth transitions can be made

5. The bases of trees and other amenities to remain

In most cases, the topographic map will have been produced through the use of photogrammetry, and most of this information will be available on the map. The engineer must determine how far beyond the limits of the project topography is required before ordering the topographic map.

Contours

Lines connecting points of equal elevation are called *contours* (Fig. 6.7). They are usually plotted for even elevations of 1, 2, or 5 m. Where the terrain is very flat, the 1 m contour interval is used and intermediate elevations are spotted where the slope between contours is not uniform. In steep terrain, the contour interval may be 5 m, 10 m, or even greater. The steeper the slope, the closer the contours will be. Therefore, rather than fill the map with contour lines, a larger interval is used.

The surveyor or photogrammetrist should have marked an elevation wherever there is a break in the slope. Therefore, it should be safe to assume that the ground between elevations slopes evenly. Though contours are used primarily to illustrate existing topographic conditions,

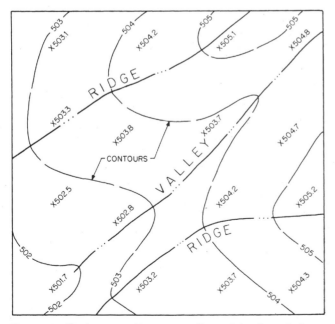

Figure 6.7 Contours are lines connecting points of equal elevation. Elevations are shown at crosses.

contour grading can be used to show proposed finished contours. During preliminary stages of design, the contours as they will exist when the construction is complete can be drawn as a graphic illustration of the concept. Exact contours can be drawn during the design phase to be used for earthwork calculations and to show drainage patterns. Contour grading is described later in this chapter.

Cross sections

Cross sections are used extensively in designing grading plans. Figure 6.8 shows examples. Elevations on the natural ground are plotted to scale in a line perpendicular to, and measured distances from, some reference line. When the points are connected, they represent the cross section of the natural ground. Then elevations at break points in the finished plan are plotted at the same horizontal location. The edge of the finished traveled way or parking lot usually does not meet the existing ground but is above or below it. This point is called the *hinge point*. From this point, a slope is designed based on the soils report. The slope will probably be between 1:1 and 1:4. That slope will be extended until it connects to the natural ground. That point is called the *catch point*.

Setting building pad elevations

The grading plan must be designed with an understanding of the drainage and roadway criteria presented in Chaps. 7 and 9. The storm drainage and overall design are coordinated with the grading plan. On hilly or complicated sites, the first step may be a preliminary contour grading plan. Usually the street profiles are designed as described in Chap. 7 and the top-of-curb elevations calculated and transferred to the grading plan. There are three types of residential lot grading plans:

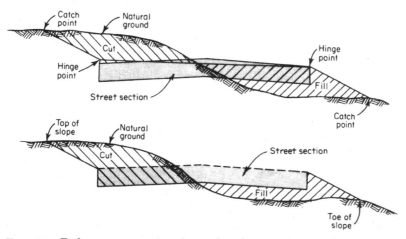

Figure 6.8   End-area cross sections for earthwork quantities.

Type A.   All the overland drainage on the lot is directed to the street at the front of the lot.

Type B.   Drainage on the front half of the lot is directed to the street in front, and drainage on the back of the lot is directed to a street, alley, or ditch in the back of the lot.

Type C.   All drainage is directed to the back of the lot.

Some jurisdictions allow only type A drainage. Where type B or C is allowed, remember that a ditch or other drainage facility must be designed for the back of the lot and that storm drainage easements must be acquired to take the drainage across adjacent property. All lots crossed with a ditch or underground system for storm drainage must be provided with a private storm drainage easement. On hillside sites where much of the site will be left natural, a ditch may be required at the property line to prevent storm water which falls on one property from crossing adjacent property.

On residential and simple commercial industrial sites, the elevations of the pads should be selected so that they will drain to the front of the property. This will save the complications of draining storm water over adjacent properties or the cost of installing storm drain inlets.

Sites with no drainage inlets.   The criteria for selecting the building pad elevations where there will be no drainage inlets within the lot are as follows:

1. The pad must be high enough above the lowest top-of-curb elevation at the front of the property to accommodate a drainage swale around the house with a slope of at least 1 percent. Often the size of the lot and slope in the street are consistent, so a constant amount can be added to the lower top of the curb to establish adjacent pad elevations.

2. The pad must be designed so the grade on the driveway does not exceed 15 percent up nor 10 percent down to the garage floor (Fig. 6.9). Steeper grades may result in the undercarriage of cars scraping and damaging the car or the driveway. Flatter driveway slopes should be used wherever possible. A drainage swale must be provided in the driveway in front of the garage where the garage is below the street.

When the building setback distance and driveway length are consistent in a subdivision, a consistent maximum elevation difference for a driveway down can be calculated. The elevation difference for a driveway up should be calculated using the top of the curb on the lower side of the driveway. The elevation difference for a driveway down should be calculated using the top of the curb on the higher side of the driveway. The driveway slope is a function of the length of the driveway as well as the elevation difference. Where flexibility is allowed for the building setback, the driveway slope can be made less steep by making the driveway longer.

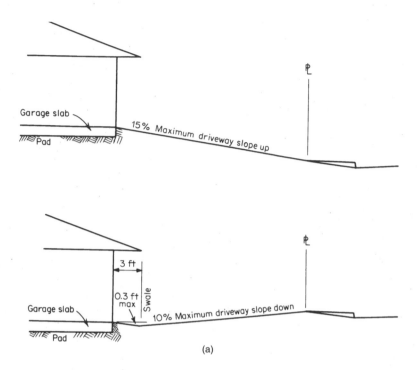

Figure 6.9 (*a*) Driveway profiles. (*b*) Car on 12 percent driveway slope.

3. The widths of slopes between pads and surrounding features are affected by the vertical distances between them. Verify that the slopes do not occupy so much space on adjacent lots that the level pad becomes too small to be useful.

4. Vertical differences between adjacent pads of less than 150 mm (0.5 ft) should be avoided. It is simpler to build three adjacent pads at one elevation and a fourth pad 180 mm (0.6 ft) different, than to build three pads each 60 mm (0.2 ft) different.

Sites with drainage inlets.   The vertical location of building pads for commercial, industrial, and multifamily residential buildings where drainage inlets will be provided must be coordinated with surrounding land:

1. The pad must be determined so that the area surrounding the pad slopes away from the building. This is to prevent storm water from entering the building.

2. The storm water release point (Chap. 9) should not be more than 300 mm (1.0 ft) above the storm water inlet.

3. The appearance of the building with respect to the street and other surroundings should be considered. If the building is much different in elevation from adjacent buildings and improvements, it will look out of place.

The size of the building pad should be designed to extend beyond the building a distance recommended by the soils engineer. Usually, the minimum distance outside the foundation to provide room to work for construction equipment and personnel is 1.5 m (5 ft). A greater distance may be required to provide for foundation support. The pad elevation should be at least 60 mm (0.2 ft) higher than is necessary to satisfy the other criteria.

Linear projects

The grading plan for linear projects will be determined initially by the horizontal and vertical alignments and will be based on criteria for those elements. When the preliminary design is complete, the cross sections or contour grading plans will be prepared and earthwork quantities calculated. The horizontal and vertical alignments can then be adjusted within the limits dictated by the horizontal and vertical criteria to effect an earthwork balance. This used to be a time-consuming task based on redrawing and remeasuring cross sections to determine earthwork quantities. Today, computer programs perform these tasks quickly, allowing the engineer to try several designs until the best fit is determined.

Adjacent property

When designing the grading plan, care should be taken at the property boundaries. All necessary earth movement and slopes must not extend beyond the property lines unless a slope or construction easement has been acquired. If the profile of a street is in cut or fill, the street must be terminated far enough inside the property to allow for the slope to reach the natural ground at the property line (Fig. 6.10). If a vertical excavation is necessary near the property line, the soils engineer should recommend treatments to stabilize the ground and protect adjacent properties and structures.

Determining Earthwork Quantities

The greatest source of earthwork is the cutting down of hills and filling in of valleys to create the pads for buildings and moderate grades for roadbeds and channels. There are three primary methods for determining earthwork quantities. Which is used depends on the situation. The three methods are averaging cut and fill depths, double end areas formula, and contour grading. Sophisticated software pro-

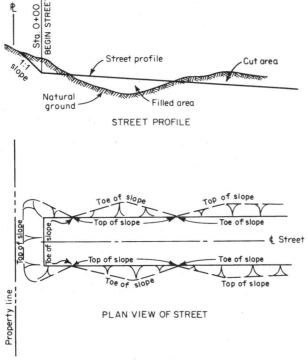

Figure 6.10   To allow for the cut slope, the street does not start at the property line.

grams are available with which the engineer can input digitized topography and roadway templates and alignments or site designs and the programs will output the earthwork quantities and the mass haul diagram. These programs are based on the following techniques. The design engineer should have a clear understanding of the basis for these programs.

Averaging cut and fill depths

The method that is the simplest, though least precise, is averaging cut and fill depths. This method is well suited to making quick, rough estimates of earthwork quantities. Its use is also indicated where the natural ground and shapes of lots are so uneven that the use of cross sections would be impractical. The method is simply to determine the difference between the existing elevation and the finished elevations and to multiply that value times the area covered.

If the site is a quarry or waste-disposal site, or any site without lots and streets, a grid is marked over the site at some convenient interval. The grid will create blocks. The size of blocks is arbitrary: 10 m (or 25 ft) each way can be used for a small site, 30 m (or 100 ft) for a large site.

For residential sites, the earthwork quantity may be calculated for each lot. The streets are then calculated separately. The finished and existing elevations at each corner are read and marked on the plan. Differences in elevations and whether to cut or fill are then marked at each corner. Where there is a hump or depression within the area of the block, it should be marked as well. The four or more points are then averaged, taking into account whether they are cut or fill. This gives the average depth, which is then multiplied by the area in square meters. This yields cubic meters of cut or fill. (If feet are used, divide the cubic feet by 27 ft³/yd³, which gives cubic yards. Where there are many blocks to calculate, a constant can be calculated representing the area divided by 27 ft³/yd³ and by the number of points averaged per lot to obtain the average depth.) Care must be taken not to use the constant where more points are used because of a hump or depression within the block.

Example 6.1 Calculate the cubic meters of cut or fill from the 30 × 30 m area shown in Fig. 6.11.

solution Calculate a constant. The square meters of one square of a grid is 30 × 30 m = 900 m². There are four elevation points per grid.

$$900 \text{ m}^2 \times \frac{1 \text{ m}}{\text{m of depth}} \times \frac{1}{4 \text{ points}} = \frac{225 \text{ m}^3}{\text{m of depth}}$$

Depths of cut are given a positive sign. Depths of fill are given a negative sign.

$$(-0.5) + (-1.4) + (0.5) + (-0.8) = -2.20 \text{ m} = \text{average depth}$$

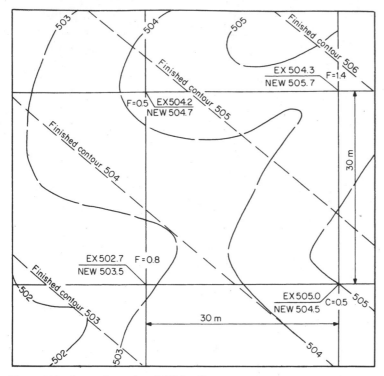

Figure 6.11   Averaging cut and fill depths.

The quantity is the grid area times the average depth.

$$\frac{225 \text{ m}^3}{\text{m}} \times -2.20 \text{ m} = 49\ 500 \text{ m}^3$$

This example uses only one square. A real task will involve many squares.

This method lends itself to use of a spreadsheet. It is desirable to show the values for each grid so that a comparison between squares can be made. This way, errors can be easily spotted. The quantities of the blocks will form a pattern or trend of increasing or decreasing values. Those blocks that do not conform to the trend should be suspect. If it is not clear from observation why those blocks do not fit the trend, the values used should be verified.

Double end areas formula

Streets, highways, and other linear structures such as canals and dikes are particularly well suited to the use of the double end area formula, which calculates the average areas of cross sections times the distances between the cross sections. The volumes are then added to yield the total volume. For design of commercial or industrial sites, the use of cross sections is also indicated.

To successfully use cross sections to calculate earthwork quantities, locations of representative cross sections must be carefully chosen and the cross sections drawn perpendicular to the centerline or other reference line (Fig. 6.12). This is true whether work is being done by hand or with a computer program.

For linear structures, cross sections should be drawn at the following locations:

1. All high and low points of the natural ground
2. Points where the profile crosses the natural ground, that is, where cut changes to fill, or fill changes to cut

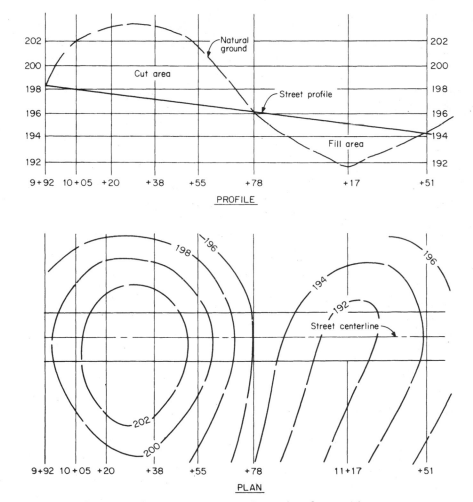

Figure 6.12 Locations of cross sections to use for earthwork quantities.

3. Points where the slope of the natural ground or the grade of the profile changes slope

4. Changes in the natural ground that are not reflected on the profile

5. Places where there are changes in appurtenant structures, such as ditches or dikes, at the sides of the main structure

For site plans, sections should be drawn at the following locations:

1. At property lines. Here the cut and/or fill should be zero, unless there is a vertical retaining wall.

2. Just before and just after changes in paved sections.

3. Just before and just after changes in building pads.

4. At high and low points in the streets or parking lots.

5. Where the topography changes.

The cross sections should be drawn to scale on grid paper (Fig. 6.13). Any scale will work; the vertical and horizontal scales may be the same, but it is not necessary. The drawing of cross sections is useful not just for calculating earthwork quantities, but as a visual aid to finding trouble spots otherwise missed and for locating cut and fill catch points (Fig. 6.8). Where the natural ground has a slope nearly the same as the designed slope for new construction, the cut or fill slope may not catch within the right-of-way or within a reasonable distance. In this case, a retaining wall may be required. However, redesign should be considered.

The cross section should include the natural or existing ground line and a template of the proposed structure. For the purpose of earthwork, the cross section can be represented in more than one way. For example, a line can be drawn across the top of the roadway. A straight line can be drawn from hinge point to hinge point or from the top of the curb to the top of the curb (Fig. 6.8a). The earthwork is then calculated as cut or fill to that line, and the area of cut below that line for the roadway structural section is subtracted.

Another way is to draw the template below the aggregate base, and calculate the earthwork from the template. Using the template provides the earthwork without the additional step of subtracting the roadway structural section (Fig. 6.8b). When representative cross sections have been drawn, the areas enclosed must be measured or calculated. The areas can be divided into trapezoids and triangles and their areas calculated, or a planimeter can be used to measure the areas. Use of a planimeter is preferable because it is faster. If the cross section is drawn on the computer, areas can be determined most easily.

Using a planimeter

The planimeter is an instrument for measuring plane surfaces. The planimeter is made up of two arms. One arm is equipped with a push-

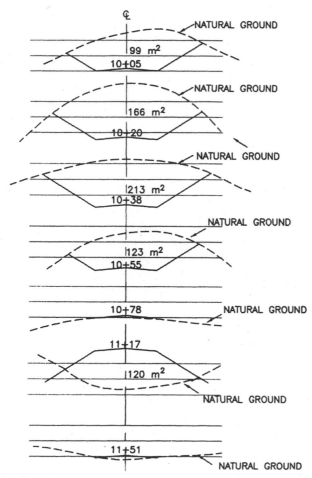

Figure 6.13 Cross sections drawn from Fig. 6.12 (imperial, not to scale).

pin at one end (to secure it to the drawing being measured) and connected to the other arm at the other end. The other arm has a pointer for tracing the area to be measured and either a graduated cylinder or a digital readout to indicate the area measured. The arms are hinged to swing freely. To use the planimeter, follow these steps:

1. Place the drawing to be measured on a flat, level surface, with no papers or other objects beneath to interrupt the smooth tracing of the area.

2. Place the static point (pushpin) so the arms form as near a 90° angle as possible when tracing the area. When the arm is extended beyond 90°, the measurement loses accuracy. Break the area into smaller areas if necessary.

3. Set the dials or readout at zero.

4. Trace the area freehand with the pointer. When tracing is done free-hand, the pointer will move inside and outside the line and the movements will compensate, but if a straight edge is used, the whole line will be inside or outside the true location, creating a cumulative error.

5. Always trace the area twice. Note the area the first time and verify that tracing the area twice yields twice the value. This is a simple check to guard against not having the dials set at zero at the beginning.

When all the cross-sectional areas have been measured or calculated and their values written on the cross sections, compare each value with the others. As a check, ask yourself, "Is the measured value of the largest area the largest number? Does the smallest visible area have the smallest measured area? Does one area that appears to be twice as large as another measure twice as large?" By examining the results critically, you may be able to discover errors. When areas have been determined for each section, the volume can be determined using the average end area formula, given in Eq. 6.1.

$$V = \frac{A_1 + A_2}{2} L \qquad (6.1)$$

where V = volume
A_1, A_2 = end areas
L = distance between end areas

This formula gives the volume of earth in cubic meters between two cross sections. (If feet are used, the volume must be divided by 27 ft^3/yd^3 to yield cubic yards.) The sum of the volumes between all the cross sections is the volume for the site. To facilitate the calculation, use the earthwork calculations form shown in Fig. 6.14. This form can easily be put into a computerized spreadsheet to facilitate the work. To fill out the form and perform calculations for the volume, follow the listed steps.

1. Write in the stations or distance from the beginning reference line in column A. There should be a station at the beginning of the project or property line, and at the end of the project or property line.

2. Enter the areas of cut in column B and the areas of fill in column F opposite their stations. The areas can be entered as square meters.

3. Determine the distances between the stations or enter the formula $(= A_2 - A_1)$ which will yield the distance between stations in column D where there is cut or column H where there is fill.

4. Add the areas of the first and second stations (column B) or the formula $(= B_1 + B_2)$ in column C between the two stations.

EARTHWORK CALCULATION FORM

(A) Station	Cut				Fill			
	(B) Area	(C) Double end Area	(D) Length	(E) Volume 1/2*(C)*(D)	(F) Area	(G) Double end Area	(H) Length	(I) Volume 1/2*(C)*(D)

Figure 6.14 Earthwork calculation form.

5. Add the areas of the second and third stations, and enter the result in column C between those two stations.

6. Continue until all adjacent areas have been added and entered on the form. The first and last stations should always have zero area unless they are at vertical retaining walls.

7. Repeat steps 4 through 6 for area in fill, filling in columns.

8. Multiply columns C and D and divide by 2. Enter the volume in column E.

9. Multiply columns G and H and divide by 2. Enter the volume in column I.

10. Total column E. The sum is the total volume of cut.

11. Total column I. The sum is the total volume of fill.

The earthwork volume for the cross sections in Figs. 6.12 and 6.13 is calculated using the form shown in Fig. 6.15. (If you are working in feet and your answer is in cubic feet, you must divide your answer by yd³/ 27 ft³. If you used a planimeter, you may have to add a factor to convert square inches to square feet.)

Contour grading

Drawing the finished contours for a proposed project is called *contour grading*. To the trained eye, contour grading provides a visual, plan-view representation of how the finished product will look and how drainage will work. Modern CADD programs for solution of earthwork volumes present the project with finished contours. These representations can be shown in plan view or with simulated aerial views to illustrate the finished product (Fig. 6.16).

EARTHWORK CALCULATION FORM

(A) Station	(B) Area	(C) Double end Area m	(D) Length	(E) Volume 1/2*(C)*(D)	(F) Area	(G) Double end Area m	(H) Length	(I) Volume 1/2*(C)*(D)
9 + 02	0							
		99	13	644				
10 + 05	99							
		265	15	1988				
10 + 20	166							
		397	18	3573				
10 + 38	231							
		354	17	3009				
10 + 55	123							
		123	23	1415				
10 + 78	0				0			
						120	39	2340
11 + 17					120			
						120	34	2040
11 + 51					0			
		Total Cut m³ =		10628		Total Fill m³ =		4380

Figure 6.15 Earthwork calculation form filled out for Fig. 6.13.

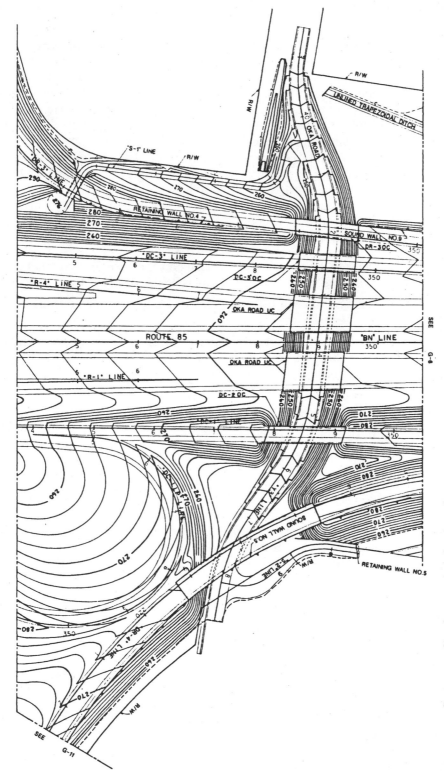

Figure 6.16 Contour grading for a portion of an interchange.

Drawing contours of a proposed project requires the ability for abstract thinking and three-dimensional visualization. One technique that can help on linear projects is to draw profiles of the break points in the cross sections. The stations at which these profiles cross even elevations can then be taken from the profile and plotted on the plan view in the correct relation to the reference line. When all the points of that elevation have been plotted, they can be connected.

Where the project is on an even grade, the contours will be at even intervals and parallel. In horizontal and vertical curves, the contours will be curved (Fig. 6.17). When drawing proposed contours on nonlinear projects, mark elevations at all break points and interpolate distances for the locations of the contours. When plotting or interpreting contours, remember the following:

1. Contour lines will always close on themselves, even though the closure may not show within the confines of the map.

2. Contour lines of different elevations never touch except in the case of a vertical wall.

3. Contour lines never cross except where there is an overhanging cliff. Here the contours under the overhang should be dashed.

4. The contours of natural ground are usually curved and seldom have abrupt changes. They should be drawn freehand.

5. On plane surfaces, as in parking lots, the contour lines are straight. If the slope is even, the contours will be parallel and evenly spaced.

6. Contour lines are perpendicular to ridge and valley lines.

The proposed contours should connect with existing contours of equal elevation at the catch points (top of slope or toe of slope). When the project is built, the new contours will start at the catch points on one side of the project, follow the proposed contours through the new area, and connect to the old contours at the catch points on the other side of the project. The old contours between catch points will cease to exist (Fig. 6.18).

Earthwork volumes can be calculated using the new contour and that portion of the old contour that will be obliterated by the project. Plane areas enclosed by the proposed and existing contours are measured with the planimeter for each elevation. Tracing the perimeter enclosed by the proposed and existing contours of each elevation in a different color helps delineate the area. A proposed contour may cross several existing contours of different elevations. (Fig. 6.19).

The volume is then calculated in the same manner and using the same form as for cross sections. Column 1 becomes contour elevations, and columns D and H are the contour intervals. When columns D and H do not change (a consistent contour interval is used) their value can be included as a constant in columns E and I.

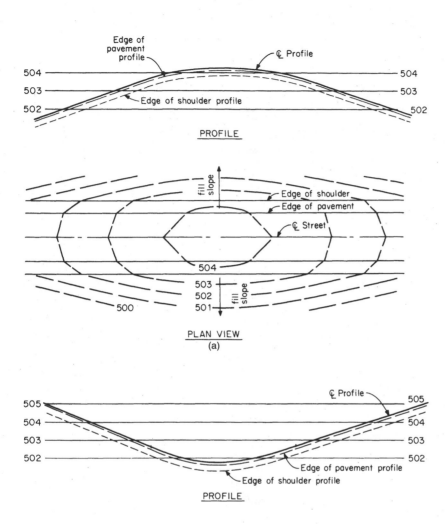

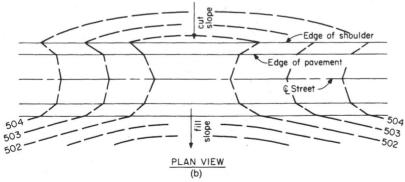

Figure 6.17 (*a*) Contour grading of crest curve. (*b*) Contour grading of sag curve.

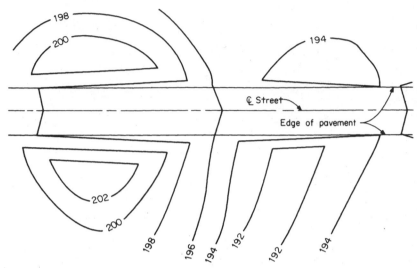

Figure 6.18 Final contours from grading Fig. 6.13.

Determining shrinkage volumes

When earth is excavated and then placed and compacted for fill, the volume will change if the degree of compaction is different before and after handling. The shrinkage factor should be given in the soils report and may differ from one part of a project to another. When calculating earthwork quantities, in order to balance the earthwork, the additional volume required to offset shrinkage must be calculated.

The relationship used to determine the amount of earth needed, compensating for shrinkage, is shown in Eq. 6.2.

$$V_R = \frac{V}{1.00 - S} \qquad (6.2)$$

where V_R = volume of compacted earth (fill) required, m³
V = volume of uncompacted earth (excavation), m³
S = shrinkage factor

Example 6.3   Determine the volume of earth needed to accommodate a shrinkage factor of 0.15. The volume of uncompacted earth is 4560 m³.

solution   Use Eq. 6.2 to determine the total volume of earth required.

$$V_R = \frac{V}{100 - S}$$

$$= \frac{4560}{1.00 - 0.15}$$

$$= 5365 \text{ m}^3$$

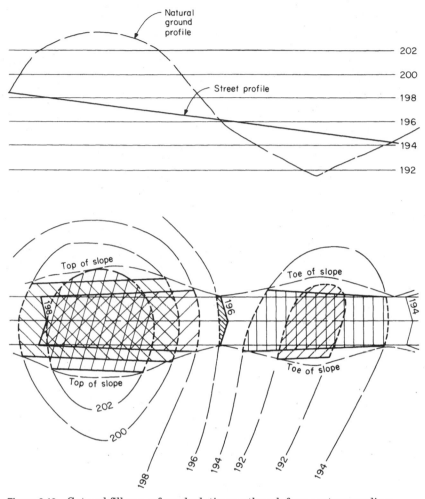

Figure 6.19 Cut and fill areas for calculating earthwork from contour grading.

Calculate the difference between the volume of compacted earth and the volume of uncompacted earth.

$$5365 \text{ m}^3 - 4560 \text{ m}^3 = 805 \text{ m}^3$$

The amount of fill to allow for the compaction factor is 805 m³.

A bulking factor should be applied to volumes of earth where the in-place material is extremely dense, such as rock. The bulking factor represents the percentage that densely compacted earth will increase in volume when it is excavated. A bulking factor is rarely needed, however, as excavated rock over 100 mm is not usually used as fill material.

Balancing the earthwork

When the cut and fill quantities have been calculated and tabulated, the total cut and fill should be compared. It is desirable for the cut and fill quantities, taking into account the shrinkage factor, to be equal. The earthwork is then said to balance. When the earthwork does not balance, all or part of the site should be raised or lowered until the site does balance.

Adjustment to cause a balance will require some redesign of grades next to the property boundaries. Earthwork quantities, including allowances for shrinkage, will have to be recalculated. This redesign and recalculation may have to be done more than once. To estimate the size of the change to be made, divide the cubic meters by the square meters of the site or area to be adjusted. The result will be the depth of material necessary to balance the earthwork spread evenly over the site. (When working in square feet and cubic yards, convert the excess or shortage of material to cubic feet by multiplying the cubic yardage by 27 ft³/yd³ then divide by the square footage of the site to yield feet of depth to adjust.)

Example 6.4   The earthwork for a site of 0.4 ha (0.98 acres) has the following earthwork tabulation:

	Cut (m³ or yd³)	Fill (m³ or yd³)
Pads and Parking	4780	2800
Compaction	—	700
Organic Material	320	—
Stockpile for Landscaping	—	320
Total	5100	3820

How much should the site be raised or lowered?

Solution

1. Calculate the difference between the cut and fill totals.

$$5100 - 3820 = 1280 \text{ m}^3$$

2. Divide the excess materials by the area of the site.

$$0.4 \text{ ha} \times \frac{10\,000 \text{ m}^2}{\text{ha}} = 4000 \text{ m}^2$$

$$\frac{1280 \text{ m}^3}{4000 \text{ m}^2} = 0.32 \text{ m}$$

At this point, redraw the cross sections 0.32 m higher, remeasure the areas, and recalculate the quantities. The cut, fill, and compaction quantities will all change. It is unlikely that the site will balance after the first redesign, because the locations of change from cut to fill will change and because the adjustment should diminish to zero at the project boundaries. With each redesign, verify that the driveway slopes and other criteria, outlined earlier, continue to be satisfied.

If the earthwork does not balance during actual construction, earth may have to be imported or exported. The cost in time and material of locating and importing earth usually exceeds the cost of exporting. Take time to plan how the site can be designed differently, that is, which pads or areas can be built higher or lower, if the site does not balance during construction. Prepare a table listing the size and location of these adjustments.

Mass diagram

Great quantities of earth are moved on linear projects. To minimize the haul distances, a mass diagram (Fig. 6.20) can be used. The *mass diagram* or *mass haul diagram* is a graph that shows where the excavation and embankment volumes are generated, the amount and volume of borrow or waste, and the direction of haul. The vertical scale or ordinate of the graph represents the algebraic sum of the quantities of excavation (a positive value) and embankment (a negative value). Shrinkage factors must be included in the calculations. The horizontal scale or abscissa represents the stations defining the length of the project. Some sophisticated earthwork software programs can produce a mass diagram while the other earthwork volumes are calculated.

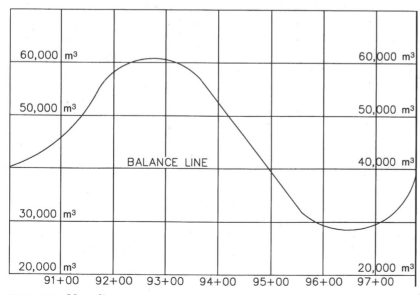

Figure 6.20 Mass diagram.

The excavation and embankment volumes are added together algebraically. Where excavation exceeds embankment, the earthwork volumes are increased. Where embankment exceeds excavation, earthwork volumes are decreased. As long as excavation exceeds embankment, the value on the chart goes up. Where the embankment material needed exceeds the excavation material, the graph turns down. A balance line is a horizontal line that is placed on the chart to indicate where the quantities balance—excavation and embankment material are the same. For the purpose of the mass diagram, the ordinate of the balance line is set at an arbitrary value which is large enough that when the embankment values are plotted, the graph will never fall below zero. The value of the balance line is added algebraically to all the excavation and embankment quantities.

The areas which are between the line of the graph and the balance line and are above it are areas where excavation is in excess of embankment. Areas between the graph and the balance line and below the balance line are areas where earth is needed for embankment. Where the graph is above the balance line, earth is transported from the left station to the station at the right at the same volume. Where the graph is below the balance line, earth is transported from the station at the right to the station at the left at the same volume.

Projects should be designed so that the earthwork will balance. Large projects will have several cycles or waves on the mass diagram. On projects where the earthwork balances, the chart will meet the balance line at the ends. However, sometimes there are political or other constraints that make a balance impossible. If the earthwork cannot be made to balance, the balance line is located where it will result in the fewest station cubic meters of haul. In some cases, best use of equipment suggests more than one balance line. In that case, one kind of equipment may be used to haul the earth long distances and a different type of equipment may be used for shorter-haul distances.

Erosion, Sediment, and Dust

When vegetation is removed from the ground, the soil beneath is vulnerable to being moved by water and wind to places that can be harmful to the environment and problems for neighbors. For this reason it is important to control the earthwork with erosion, sediment, and dust control measures.

One method of control employed by some jurisdictions is to disallow grading after some specified date in the fall and before some specified date in the spring. When grading is allowed, check dams and sedimentation basins may be required. When grading is taking place, the roadways being used to transport the earth are kept watered or treated with some other dust palliative. When grading is complete, the new slopes are planted with native plants or landscaped to ensure that erosion does not harm the slopes and pollute surrounding area streams and waterways.

In 1987, an amendment to the Clean Water Act (CWA) of 1972 established a requirement that discharges of storm water associated with construction activities should be regulated. Thereafter, wherever the area of construction activities cover more than 2 ha (5 ac), the work is treated as an industrial activity and requires a National Pollution Discharge Elimination System (NPDES) permit. The purpose of the permit is to prevent soil or construction materials from reaching waterways and polluting them.

To acquire the permit, a Storm Water Pollution Prevention Plan (SWPPP) and a notice of intent (NOI) must be prepared and submitted. The notice of intent is simply a form that must be filled out stating location of the project site, when work will begin and end, location of the receiving waters, who owns the site, and who will be responsible for construction activities. The Storm Water Pollution Prevention Plan is a description of what measures will be taken to prevent any undesirable substance from reaching storm water inlets or waterways. The methods employed are sheds or plastic sheeting to contain hazardous materials such as paints and oils, and earth berms, hay bales, silt fences, and detention (settling) ponds to stop silt from reaching waterways. These techniques are called *best management practices* (BMP) and local storm water agencies should have detailed descriptions of them. A plan showing where the locations of various construction activities such as storage of waste products, construction equipment, and hazardous materials must be drawn and included in the SWPPP.

A program to educate construction workers as to the importance of the best management practices and responsibilities must be described. Further, a maintenance program and schedule and statement of who will be responsible for containment in case of a serious spill must be spelled out. Failure to get the permit or to comply with the provisions of the plan can result in serious penalties and even jail time for the owners and other responsible persons. Most of the information for the permit is best prepared by the construction superintendent but, because a drawing is required, the site design engineer may be asked to prepare the plan.

The Grading Plan

All the information the grading contractor will need should be shown on the grading plan (Fig. 6.21). It should be drafted in such a way as to be completely independent of the rest of the plans for the site. The grading plan base map should be made on a topography map and drawn to scale. It is useful to have the topography map screened in plotting or photographically to show proposed activities clearly.

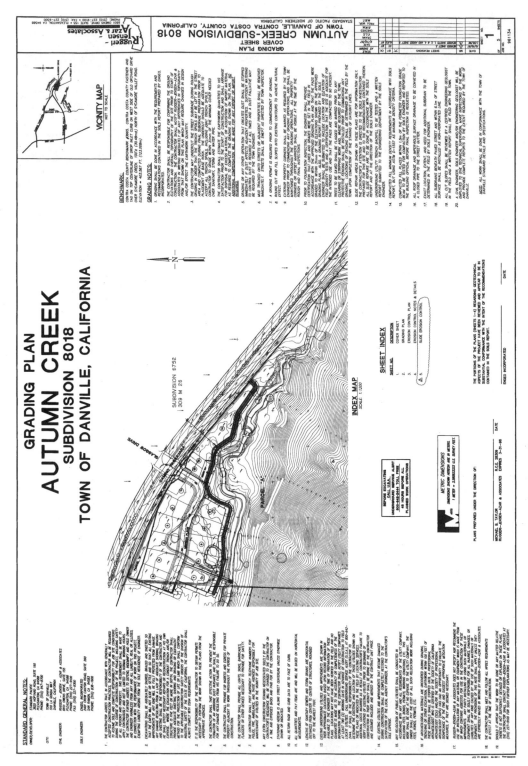

Figure 6.21 Grading plan. (*Courtesy of Rugeri, Jensen, Azar & Associates.*)

138

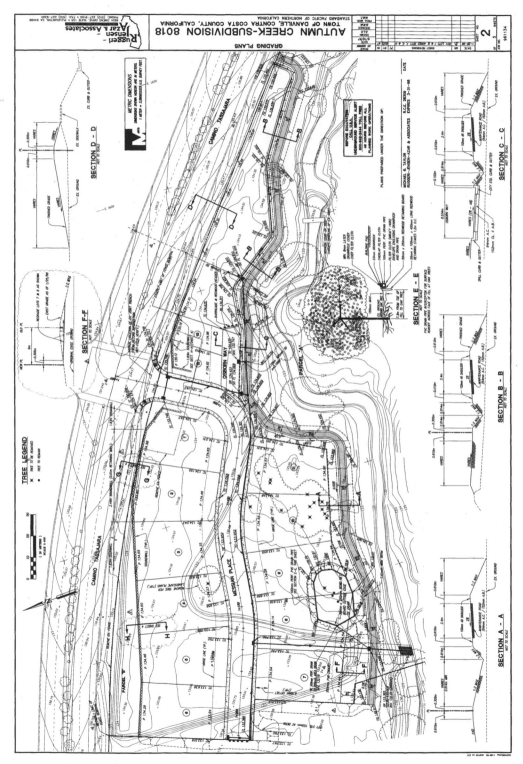

Figure 6.21 (*Continued*)

The property boundary, lot, and easement lines are then drawn on the screened copy. This way the topographic information is available and serves as a background on which to present the plan. The plan should show a typical section or sections for the pads and for the streets with slopes and setback distances from property lines and structures. Cross sections for ditches, dikes, and other drainage structures that are an integral part of the plan and will be the responsibility of the grading contractor should be shown. On linear projects, a mass diagram may be a part of the grading plan set. Grading plans for site projects should show the areas of fill shaded or otherwise delineated differently from areas of cut, to show the extent of transporting earth around the site.

An important part of the plan is the notes. The notes should include the names of the project engineer, the soils engineer, and the jurisdictional agency. A table of the cut and fill quantities and a table of alternative locations for adjustments may be included. Instructions on how to handle special problems, such as underground tanks, should be given. If erosion control measures are to be taken, they should be described in the notes and details. A checklist for grading plans is shown as Fig. 6.22.

Summary

Earthwork is one of the major aspects and sources of costs in construction. The soils and geologic reports provide information about the characteristics of the site and the soils. Some of that information determines design factors and information for calculating earthwork quantities. There are a number of sources of earthwork and three basic ways of determining what the earthwork quantities will be. Because of its sig-

---

GRADING PLAN CHECKLIST

a.	Existing topography
b.	Vicinity map
c.	North arrow
d.	Property boundaries; bearings and distances
e.	Lots and blocks or parcels, easements; distances
f.	Street names
g.	Bench mark
h.	Title block
i.	Typical sections
j.	Top of curb elevations
k.	Pad elevations. Areas of cut or fill should be shaded on large projects.
l.	Slopes
m.	Ditches
n.	Bulkheads or retaining walls
o.	Details
p.	Notes

Figure 6.22 Grading plan checklist.

nificance to costs, the engineer must have a clear understanding of the factors involved and how to estimate earthwork quantities accurately.

Problems

1. Name three things that should be included in the geologic report.

2. Name three things that should be included in the soils report.

3. What is liquefaction?

4. What is shrinkage?

5. Define earthwork balance.

6. What is the greatest source of earthwork?

7. What is a benchmark?

8. What is a benchmark equation?

9. Define contour.

10. What are the three types of lot grading plans?

11. Describe three methods of determining earthwork volumes.

12. What is a catch point?

13. What is a hinge point?

14. Describe two reasons to make contour grading plans.

15. What volume of excavated material is needed to fill 6027 m³ of embankment? The shrinkage factor is 15 percent.

16. The earthwork for a site has the following earthwork tabulation. The site covers 2 ha (5 ac). How much should the site be raised or lowered?

	Cut (m³ or yd³)	Fill (m³ or yd³)
Pads and Parking	7440	11 000
Compaction	—	1100
Organic Material	2020	—
Stockpile for Landscaping	—	2020
Total	9460	14 230

17. Define SWPPP.

Further Reading

Association of Bay Area Governments, *Manual of Standards for Erosion and Sediment Control Measures,* Berkeley, California, 1981.

Bay Area Stormwater Management Agencies Association, *Blueprint for a Clean Bay,* Oakland, California, 1995.

Intergraph Corporation, *Inroads,* Huntsville, Alabama, 1992.

Softdesk, *Civil Engineering and Surveying,* Henniker, New Hampshire, 1992.

State Water Resources Control Board, *Fact Sheet for National Pollutant Discharge Elimination System (NPDES) General Permit For Storm Water Discharges Associated With Construction Activity,* Sacramento, California, 1992.

Roadways and Parking Lots

In designing roadways, the primary consideration is to provide a safe, smooth, and comfortable ride. This is true whether the traveled way is a major highway or a neighborhood street. The primary difference in design considerations between rural roadways and city streets is that city streets are typically an integral part of the storm drainage system. Though storm drainage is an important consideration that must be planned for on rural roadways, it has far less influence on design.

Horizontal Alignment

In open country, the horizontal location of a roadway is determined by the shortest route between two points, the alignment that requires the least earthwork, the stopping and passing sight distances, and the preservation of natural resources. In addition, traffic patterns, the value of property, and the location of established neighborhoods influence the location of highways in urban areas. In locating a street in new subdivisions, the way that street divides the property relative to the lots it delineates is the dominant criterion.

Establishing the reference line

Streets and roadways must have a reference line from which their various elements can be located. On simple streets, the centerline is ordinarily used. On roads where the opposing traveled ways are separated and on one-way ramps, the *axis of rotation* (Fig. 7.21) is usually most convenient. The axis of rotation is the point on the cross section about which the cross section is turned. As long as the location of the reference line is clear, the reference line can be anywhere. For simplicity, the centerline will be the reference line in this book.

The location of the centerline of a roadway must be established relative to its legal environment. The way to accomplish this is with a bearing and a distance from a known point. For a highway, it is usually a named station of an existing highway reference line. In a subdivision, the location is tied to the property line from a property corner. From there, the courses of the reference line are described with bearings, distances, and curve data. The reference line is tied to other known points where it crosses property lines or streets. Where centerlines cross, stationing equations, described later, are given.

Where the centerline changes direction, a tangent, circular curve is inserted for a smooth transition. In some states, spiral curves are added at the ends of the circular curves on highways. Spiral curves are also used in the design of railroads. Only circular curves will be described in this book.

The geometric relationships of circular curves are illustrated in Fig. 7.1. The radius, the central angle, and the length are all that is necessary to describe a circular curve. This information should be shown on the plans, either next to the curve or in a table. The radius and deflection or central angle (Δ) are usually known or are set during design. When the radius or deflection is not known, the equations given in Fig. 7.1 can be manipulated algebraically to give a value for either one. Most computer programs simply require that you input whether the

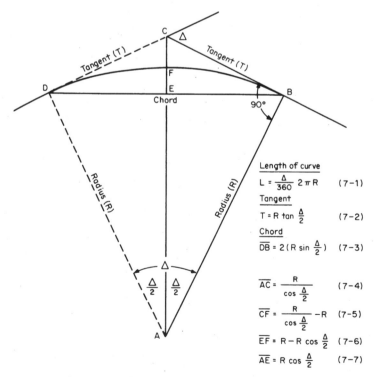

Length of curve

$$L = \frac{\Delta}{360}\, 2\pi R \qquad (7\text{-}1)$$

Tangent

$$T = R \tan \frac{\Delta}{2} \qquad (7\text{-}2)$$

Chord

$$\overline{DB} = 2\left(R \sin \frac{\Delta}{2}\right) \qquad (7\text{-}3)$$

$$\overline{AC} = \frac{R}{\cos \frac{\Delta}{2}} \qquad (7\text{-}4)$$

$$\overline{CF} = \frac{R}{\cos \frac{\Delta}{2}} - R \qquad (7\text{-}5)$$

$$\overline{EF} = R - R \cos \frac{\Delta}{2} \qquad (7\text{-}6)$$

$$\overline{AE} = R \cos \frac{\Delta}{2} \qquad (7\text{-}7)$$

Figure 7.1 Geometric relationships of horizontal curves.

curve is to the right or left, the radius, and the length of the curve or the central angle.

Stationing

The reference line is stationed. *Stationing* means simply marking the distance every 100 m (100 ft in the English system). The stations are numbered consecutively starting with 0 at some convenient location, such as the county line for highways and a tract boundary or street intersections for subdivisions. The stations are labeled as STA 1+00, STA 2+00, STA 3+00. . . . Where a station or a measurement is needed at some point other than an even 100 m (100 ft) from the previous station, the station is given as station plus meters; thus, STA 1+32.66 is 132.66 m (132.66 ft in the English system) from STA 0+00. When using stations for calculations, show stations in parentheses. The + in the station is not a mathematical operation.

Whenever possible, it is desirable for the stationing to run from left to right on the plans. The bearings and distances of the centerline are calculated and then labeled in degrees, minutes, and seconds, and in meters, respectively. Wherever there are curves, the radii, central angles (Δ), and lengths of the curves must be calculated. This information is all that is needed to station the roadways. This is illustrated in Fig. 7.2. Here the centerline distance from the tract boundary (STA 0+00) to the beginning of the curve (BC) is calculated to be 162.03 m, the length of the curve is 329.87 m, and the centerline distance from the end of the curve (EC) to the other tract boundary is calculated to be

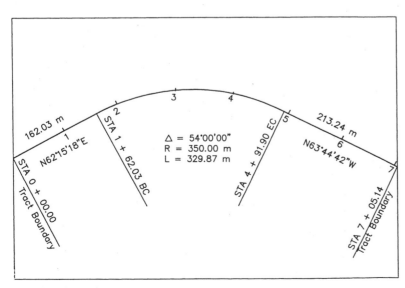

Figure 7.2   Stationing a curve.

213.24 m. The BC station is at STA 1+62.03, the EC is at STA 4+91.90, and the tract boundary is at STA 7+05.14.

In some circumstances, a stationing equation is needed—for example, when northbound and southbound lanes are separated by a median and the distance between the lanes varies so that the traveled ways are not parallel. This situation is not unusual in mountainous areas. The highway may be designed this way so that separate profiles are needed for hundreds of meters or even many kilometers; then it returns to a parallel design. Where the traveled ways leave and return to a parallel condition, stationing equations are needed.

A stationing equation might also be needed where portions of an old road are redesigned so that the stationing changes back and forth between old and new sections.

Example 7.1   Determine the stationing equation at the EC of the new centerline of Winding Road as illustrated in Fig. 7.3. Use the equations in Fig. 7.1. (For the purpose of this example, it does not matter whether feet or meters are used.)

solution

1. We see from the illustration that the centerlines coincide except for the portion within the curves. Therefore, we know that the central angle of each curve is the same. We know from geometry that curves with different radii and the same central angle have different curve lengths and tangent lengths.

2. Calculate the tangent length of the existing curve and the new curve using $T = R \tan (\Delta/2)$ (Eq. 7.2). First calculate $\Delta/2$.

$$\frac{\Delta}{2} = \frac{83° \, 40'}{2}$$

$$83° = 82° \, 60'$$

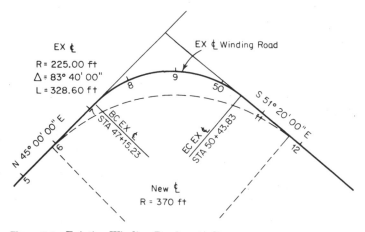

Figure 7.3   Existing Winding Road centerline.

$$82° + \quad 60'$$

$$\frac{+ \quad 40'}{82° \quad 100'}$$

$$\frac{82°}{2} + \frac{100'}{2} = 41°\ 50'$$

$$\frac{83°\ 40'}{2} = 41°\ 50'$$

The tangent of 41° 50′ is 0.8951506.

$$T_{\text{new}} = 370 \text{ m} \times 0.8951506 = 331.21 \text{ m}$$

$$T_{\text{existing}} = 225 \text{ m} \times 0.8951506 = 201.41 \text{ m}$$

3. Calculate the difference of tangent lengths.

$$331.21 \text{ m} - 201.41 \text{ m} = 129.80 \text{ m}$$

4. Subtract the difference of tangent lengths (129.80 m) from the BC station of the existing centerline. Remember that, in equations, station numbers are given in parentheses to distinguish them from lengths. The + is not a mathematical operation.

$$(\text{STA } 47+15.23) - 129.80 \text{ m} = (\text{STA } 45+85.43)$$

This is the BC station of the new centerline.

5. Calculate the length of the curve of the new centerline. Use Eq. 7.1 in Fig. 7.1. First convert degrees, minutes, and seconds to degrees and decimals of degrees.

$$\frac{40'}{60'/\text{deg}} = 0.667°$$

$$83°\ 40' = 83.67°$$

$$L = \frac{\Delta}{360}\ 2\,\pi\,R$$

$$L = \frac{83.67°}{360} \times 2 \times \pi \times 370 \text{ m}$$

$$L = 540.04 \text{ m}$$

6. Determine the station of the new EC.

$$\text{New BC STA + length of new curve = new EC STA}$$

$$(\text{STA } 45+85.43) + 540.04 \text{ m} = \text{STA } 51+25.47$$

7. Determine where the new EC becomes tangent to the old centerline. Add the tangent difference of 129.80 m to the EC of the existing centerline.

$$(STA\ 50+43.83) + 129.80\ m = (STA\ 51+73.63)$$

The equation is

$$EC\ New\ \text{\textcentoldstyle}\ STA\ 51+25.75\ NEW = POT\ EX\ \text{\textcentoldstyle}\ STA\ 51+73.63$$

The new centerline will be 48.14 m shorter than the old one (Fig. 7.4).

When you calculate slopes along conduits or along curbs in curves, you should be careful not to use the difference of stationing as the length of conduits or curbs. The length of the curb must be determined from the plan view. The length of a curb will be shorter than the centerline when it is concentric with and inside the centerline curve. It will be longer when it is concentric with and outside the centerline curve.

This fact is important when designing drainage where minimal slopes are needed. The curb that is concentric with and outside the centerline is longer than the centerline; therefore, the slope on it will be flatter. The slope on the outside curb will be set at the minimum and will determine the minimum acceptable slope for the centerline. If the work is not being done on computer, a simple formula can be used for calculating the length of a curve. The formula for the curve which is not on the centerline is the same as it is for the centerline (Eq. 7.1). The only value that is different is the radius.

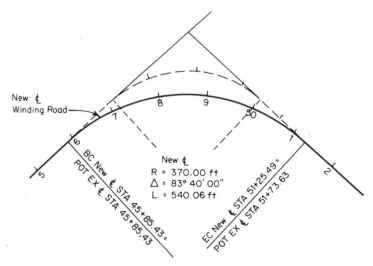

Figure 7.4 Proposed Winding Road centerline.

$$L_{\textcent} = \frac{\Delta}{360°} \, 2 \, \pi \, R_{\textcent} \qquad L_{\mathrm{CRB}} = \frac{\Delta}{360°} \, 2 \, \pi \, R_{\mathrm{CRB}}$$

$$\frac{L_{\textcent}}{R_{\textcent}} = \frac{\Delta}{360°} \, 2 \, \pi \qquad \frac{L_{\mathrm{CRB}}}{R_{\mathrm{CRB}}} = \frac{\Delta}{360°} \, 2 \, \pi$$

$$\frac{L_{\textcent}}{R_{\textcent}} = \frac{L_{\mathrm{CRB}}}{R_{\mathrm{CRB}}}$$

A constant can be calculated for the relationship

$$\frac{L_{\textcent}}{R_{\textcent}}$$

This constant can then be used to calculate the lengths of all other lines that are concentric with the centerline.

$$\frac{L_{\textcent}}{R_{\textcent}} = K \tag{7.8}$$

$$K = \frac{L_{\mathrm{CRB}}}{R_{\mathrm{CRB}}} \tag{7.9}$$

$$KR_{\mathrm{CRB}} = L_{\mathrm{CRB}} \tag{7.10}$$

Example 7.2   Calculate the length of the right curb between STA 5+00 and STA 8+92.70 from Fig. 7.5.

solution

1. Using Eq. 7.8, calculate a constant for the relationship between the centerline radius and the length of the centerline curve. (For the purpose of this example, it does not matter whether feet or meters are used.)

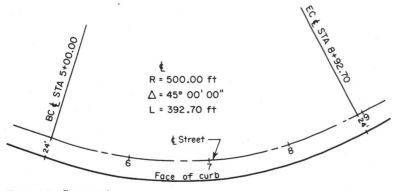

Figure 7.5   Concentric curves.

$$L_{\mathfrak{C}} = \frac{392.70 \text{ m}}{500.00 \text{ m}} = 0.7854 = K$$

2. Calculate the radius of the right curb.

$$500.00 \text{ m} + 24.00 \text{ m} = 524.00 \text{ m}$$

3. Using Eq. 7.10, calculate the length of the right curb.

$$L_{\text{CRB}} = K \times R_{\text{CRB}}$$

$$= 0.7854 \times 524.00 \text{ m}$$

$$= 411.55 \text{ m}$$

Horizontal sight distances

The distance required for drivers to bring a car traveling at a given speed to a stop after sighting an object in the road is called the *stopping sight distance*. On a horizontal curve, the stopping sight distance is measured as the line of sight (a straight line) from one point on the curve to another. The line of sight forms a chord of the curve. The area enclosed within the curve and the line of sight must be kept clear of obstructions (Fig. 7.6). The approving agency will dictate minimum radii to be used for curves and distances from the curves for which the line of sight must be preserved.

Truck turns

On projects where truck traffic is expected and small-radius curves are used at curb returns, the lanes must be wider to accommodate truck turns. Long trucks require wider traveled ways for turning because the rear tires do not track the front tires. The paths that the front and rear tires take have been established through empirical methods. Templates

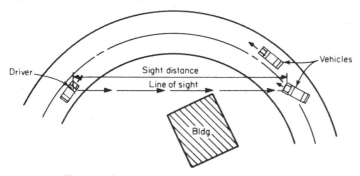

Figure 7.6 Horizontal stopping sight distance.

showing the paths of the tires can be acquired from the Federal Highway Administration (Fig. 7.7). The templates should be used to determine the pavement widths wherever short-radius curves are used.

Vertical Alignment

The drawing used for the longitudinal, vertical design of roadways and utilities is called a *profile*. A profile shows the natural ground and the roadway profile at the centerline. The profiles of sewers and other utilities are projected into the vertical plane of the centerline. Other information necessary or helpful for designing the profiles should also be shown. Crossing utilities and other traveled ways are particularly important. The stationing of the reference line is shown on the profile. It is not affected by the slope of the profile but is taken from the plan view. When calculating slopes for utilities, care must be taken not to use the difference of stationing as the length of the conduit or curb when the centerline alignment is a curve.

Designing the roadway profile

The first step in designing a profile is drawing it graphically. The existing ground and all critical crossings and other critical points for the entire length of the profile should be plotted in a continuous profile.

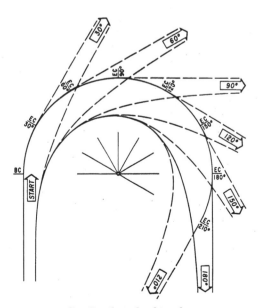

Figure 7.7   Semitrailer wheel tracks.

Select the horizontal and vertical scales so that the entire profile can be shown without breaks. Draw the natural ground at the centerline, and show the existing profiles of the roadways to be connected to or crossed. Features that may affect the vertical alignment, such as other roadways or bridges to be crossed or connected to, existing underground utilities, trees to be saved within the right-of-way, or overhead power lines, should be plotted and their elevations shown.

In subdivisions where adjacent lots must be drained to the street, the profile should be set low enough to allow drainage from the lots. The existing natural ground may not be a factor in that situation. On lots 30 m (100 ft) from front to back and 15 m (50 ft) wide, the overland flow will have to travel half the width plus the length of the lot

$$\frac{15}{2} + 30 = 37.5 \text{ m}$$

At 1 percent fall for 37.5 m (123 ft), the street should be located vertically so that the highest top-of-curb elevation at the property line is at least 0.375 m (1.23 ft) below the pad elevations (37.5 m × 0.01).

Balancing earthwork is always an important criterion. The roadway profile should follow the slope of the natural ground as much as possible while taking into account the depth of the structural section. The more closely it follows the ground, the less earthwork will be required. Where changes occur in the natural grade, a change should be made in the slope of the roadway profile. By putting your eye next to the paper at one end of the profile, the location of the angle point in the slope of the natural ground will become apparent.

Judgment must be exercised as to how closely to follow the natural ground. Too many grade breaks result in an unsafe, uncomfortable ride and increased engineering and surveying costs. The profile must allow enough vertical clearance to allow for culverts or other drainage facilities and must be coordinated with other existing and proposed surface, subsurface, and overhead conditions and improvements. In hilly terrain, the natural ground will be used only as a guide to locating the profile. Balancing the amounts of excavation and embankment is a more important consideration.

Profile slopes

Consideration must be given to limiting the steepness of slopes (grades) used for roadways. The approving agency will have minimum and maximum slope limits. The maximum slope to be used for long distances of highway may be as low as 6 percent. This slope is chosen so that a heavily laden truck can be expected to climb or descend safely. City streets may be designed with grades as steep as 15 percent for short distances. For private roads, 20 percent may be used if allowed by local agencies. Here, access for emergency vehicles, such as fire trucks, is a consideration. Where roadways have slopes of 10 percent or greater, it may be wise to limit parking. The force of gravity becomes a factor in handling car doors where the slope of the parking area is steep. An unwieldy door can be a hazard for small adults or for children.

When a roadway is in open country and can be built above the surrounding land, the road can drain laterally and the profile can actually be level. When streets are designed in subdivisions, they may be used to channel and control storm water and must be designed with longitudinal slopes sufficient to allow drainage. A slope of 0.003 is an absolute minimum. Successful construction of a slope that flat is questionable, and bird baths (small puddles) are likely to develop. Use a steeper slope whenever possible. The approving agency will dictate the minimum allowable slope.

Calculating the profile

Once a profile for a roadway has been designed graphically, its location must be calculated. The slope is calculated by dividing the horizontal difference between break points into the vertical (elevation) difference between the same points. These points can be measured from the graphic layout of the profile or, where there is a critical point, a known elevation.

Example 7.3   Calculate the slope between STA 10+00 and STA 18+00 from Fig. 7.8. (For the purpose of this example, it does not matter whether feet or meters are used.)

solution

1. Calculate the distance between STA 10+00 and STA 18+00.

$$(STA\ 18{+}00) - (STA\ 10{+}00) = 800\ m$$

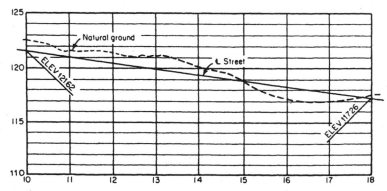

Figure 7.8   Example 7.4. Slope of profile.

2. Calculate the difference in elevation between STA 10+00 and STA 18+00.

$$121.62 \text{ m} - 117.26 \text{ m} = 4.36 \text{ m}$$

3. The slope is the elevation difference divided by the distance between stations.

The slope is 0.00545.

Often engineers express surface slopes as percentages and subsurface slopes on pipes as 1 per 100. The surface slope of 0.545 percent is expressed as 0.00545 on subsurface structures. The slope in example 7.4 is 0.545 percent and is said to be negative because it is going downhill as the stationing increases. This process can also be reversed. When the slope and the elevation at one station are available, the unknown elevation at another station can be calculated.

Example 7.4   Given an elevation of 121.62 at STA 10+00 and a slope of −0.00545, calculate the elevation at STA 15+30.54. (For the purpose of this example, it does not matter whether feet or meters are used.)

solution

1. Calculate the distance between stations.

$$(\text{STA } 15+30.54) - (\text{STA } 10+00) = 530.54 \text{ m}$$

2. Multiply the slope times the distance.

$$-0.00545 \times 530.54 \text{ m} = -2.89 \text{ m}$$

This gives the vertical difference in elevations.

3. To get the elevation at STA 15+30.54, add the difference algebraically to the elevation at STA 10+00.

$$121.62 + (-2.89 \text{ m}) = 118.73$$

The elevation at STA 15+30.54 is 118.73.

When the roadway slope changes more than 2 percent, a vertical curve should be used. The length of the vertical curve to be used is determined by passing and stopping sight distances and what curve best fits other criteria. These are discussed later. The vertical curves used for roadways, and sometimes sewers, are parabolic curves. Deriving the formulas for parabolic curves requires an understanding of calculus. Fortunately, all that is needed to calculate and plot vertical curves are three simple equations (Fig. 7.9).

When the profile goes from a positive (uphill) slope to a negative (downhill) slope, from a positive slope to a flatter positive slope, or from a negative slope to a steeper negative slope, the curve used is said to be a *crest curve* (Fig. 7.10). When the profile goes from a negative slope to a positive slope, from a negative slope to a flatter negative slope, or

Point of intersection

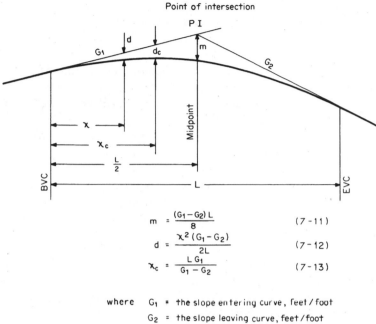

$$m = \frac{(G_1 - G_2) L}{8} \qquad (7-11)$$

$$d = \frac{x^2 (G_1 - G_2)}{2L} \qquad (7-12)$$

$$x_c = \frac{L G_1}{G_1 - G_2} \qquad (7-13)$$

where G_1 = the slope entering curve, feet / foot

G_2 = the slope leaving curve, feet / foot

L = length of curve, ft

x = any horizontal distance on curve

x_c = horizontal distance to critical point from the near end of curve

m = vertical offset distance at midpoint

d = vertical offset distance

d_c = vertical offset distance at critical point

Figure 7.9 Geometrics of vertical curves.

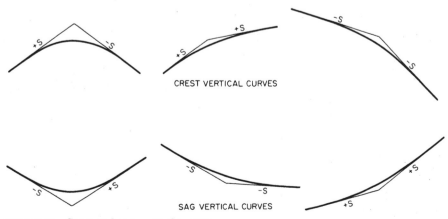

CREST VERTICAL CURVES

SAG VERTICAL CURVES

Figure 7.10 Crest and sag vertical curves.

from a positive slope to a steeper positive slope, the curve is called a *sag curve*. The highest point on a crest curve and the lowest point on a sag curve are called the *critical points*. Sag curves form bowls. Crest curves form inverted bowls.

When the slopes are equal in value but opposite in direction, the critical point falls at the station of the point of intersection (PI) of the tangent slopes. When grades change but both are positive or both are negative, the critical point falls at one end. In all other situations, that is, where the slopes are different in directions, one positive and the other negative, and different in value, the critical point falls within the curve. In such cases, the horizontal location of the critical point must be calculated first—then the vertical elevation calculated using the horizontal location. The critical point on sag curves is particularly important because it determines where drainage facilities must be located.

To lay out smooth vertical curves, elevations should be calculated at the critical point, at the midpoint, and at 10 m (25-ft) intervals. Plotting the curve on the profile sheet is done with french curves. The curve should be tangent at the beginning of the vertical curve (BVC) and at the end of the vertical curve (EVC) and should go through three elevation points.

Example 7.5   From Figs. 7.9 and 7.11, calculate the elevation at the midpoint of the curve and the location and elevation at which to place catch basins between STA 23+20 and STA 27+00. Use a 320 m vertical curve. (For the purpose of this example, it does not matter whether feet or meters are used.)

solution

1. The center of the curve will be at the PI, STA 25+00. The BVC will be located back of the PI by half the length of the vertical curve.

$$\frac{320 \text{ m}}{2} = 160 \text{ m}$$

BVC STA = (STA 25+00) – 160 m = (STA 23+40)

EVC STA = (STA 23+40) + 320 m = (STA 26+60)

2. Calculate the elevation at the PI.
   a. Determine the distance between a station of known elevation STA 23+20 and the PI station.

   (STA 25+00) – (STA 23+20) = 180 m

   b. Multiply the distance by the slope.

   180 m × (–0.05) = –9.00 m

   c. Add algebraically the vertical difference to the known elevation.

   276.32 m + (–9.00 m) = 267.32 m

3. Calculate the vertical offset distance at the midpoint.

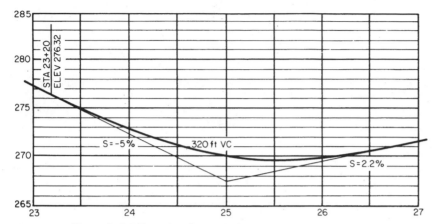

Figure 7.11   Example 7.6. Locating the critical point.

$$m = \frac{(G_1 - G_2)\,L}{8}$$

$$= \frac{(-0.05 - 0.022)\,(320\text{ m})}{8} \qquad (7.11)$$

$$= 2.88 \text{ m}$$

This is a sag curve, so the elevation at the midpoint of the curve is the elevation at the PI plus the distance m. On crest curves, the distance m is subtracted from the PI.

$$267.32 \text{ m} + 2.88 \text{ m} = 270.20 \text{ m}$$

The elevation on the curve at the midpoint is 270.20.

4. Calculate the location and elevation of the critical point.
   a. Determine the distance from the *near* end of the curve to the critical point. The approximate location of the critical point should be apparent visually. It will fall between the PI and the end of the curve where the slope is flatter—not on the steeper side. The value will be less than half the length of curve. It can be seen from Fig. 7.11 that the critical point will lie between STA 25+00 and the end of the vertical curve (EVC). To calculate the critical point in that area, G_1 will be 0.022 (2.2 percent) and will be negative as viewed from the end of the curve nearest the critical point—in this case the EVC. G_2 will be 0.05 (5 percent) and will be positive as viewed from the end of the curve nearest the critical point. Notice that in Fig. 7.9 the critical point is near the BVC. Use Eq. 7.13.

$$X_c = \frac{L\,G_1}{G_1 - G_2}$$

$$= \frac{320\,(-0.022)}{-0.022 - 0.05}$$

$$= 97.78 \text{ m}$$

The critical point is 97.78 m from the EVC. The calculation of the critical point is

$$\text{EVC (STA 26+60)} - 97.78 \text{ m} = \text{STA 25+62.22}$$

b. Calculate the elevation at the critical point. The elevation on the tangent must be calculated first.

$$(\text{STA 25+62.22}) - (\text{STA 25+00}) = 62.22 \text{ m}$$

$$267.32 \text{ m} + (62.22 \text{ m} \times 0.022) = 268.69 \text{ m}$$

The offset distance d_c is calculated using Eq. 7.12. Since the offset distance d_c is being calculated, X_c is used as the horizontal distance X.

$$d_c = \frac{X^2 (G_1 - G_2)}{2 L}$$

$$= \frac{(97.78)^2 (-0.05 - 0.022)}{2 \times 320 \text{ m}}$$

$$= 1.08 \text{ m}$$

The vertical offset distance must be added (on a sag curve) to the elevation on the tangent at the station of the critical point.

$$268.69 \text{ m} + 1.08 \text{ m} = 269.77 \text{ m}$$

The catch basin should be located at STA 25+62.22, where the elevation is 269.77. The elevation at the top of the grate (TG) of the catch basin will be somewhat lower because of the cross slope of the roadway. If the horizontal alignment is curving, the location may shift because of the cross slope being affected by a superelevation. A description of superelevation follows.

Meeting existing roadways

The elevation of the point on the existing roadway where the new roadway begins should be known from the topographic survey. It is not sufficient to simply use the elevations of centerlines and tops of curbs-to-be-met from existing plans. A survey crew must locate existing centerline elevations to within 0.03 m vertically at the point of intersection of the proposed and existing roadways and at 10 m (25-ft) intervals and at grade breaks back along the existing roadway for 15 m (50 ft) or more behind the connection. This information will provide the existing slopes. The approving agency will dictate the maximum grade break allowed without a vertical curve. If they do not have a policy and the algebraic change of slope between the new and existing roadway is greater than 2 percent, a vertical curve should be used. The point of intersection (PI) of the new and the existing slopes should be moved into the new profile a sufficient distance so that the vertical curve selected fits wholly in the new pavement if possible. If that approach is not practical, the removal of some of the existing roadway may be required. Experiment with the location of the point of intersection to reduce the amount of existing pavement to be removed.

Vertical sight distances

The distance a driver can see ahead on the road is called the *sight distance*. The distance along the road that a driver must be able to see ahead to safely pass a car traveling at a particular speed is called the *passing sight distance*. The distance it takes a driver, traveling at a particular speed, to stop a car after seeing an object in the road is called the *stopping sight distance* and is different for crest and sag curves. Streets and roads are designed to allow traffic to travel at particular speeds. The speed of traffic determines the necessary stopping and passing sight distances. The desirable stopping or passing sight distances dictate the lengths of vertical curves necessary to provide safe driving through grade changes. Figure 7.12 can be used to select the appropriate stopping sight distance on a crest vertical curve. Figure 7.13 shows curve lengths for stopping sight distances on a sag curve.

Example 7.6   Determine the length to use for a vertical curve when the entering slope is +1 percent and the exiting slope is +5 percent. The design speed is 80 km/h.

solution

1. The curve is a sag curve because the slope goes from a positive slope to steeper positive slope.

2. The algebraic difference in grades is

$$(+5) - (+1) = 4 \text{ percent}$$

3. Enter the chart for stopping sight distance in a sag curve (Fig. 7.13) at 4 percent on the left side. Follow the 4 percent line to its intersection with the 80 km/h line. The length of the curve to use is 116 m, as read directly at the intersection.

Profile calculations form

Calculations for the profile should be recorded in an orderly manner. A profile calculations form is given as Fig. 7.14. An example of the completed form is shown in Fig. 7.15. Notice that a profile with a station equation is illustrated. Because of the equation, the actual distance between STA 51+00 and 52+00 is 52.17 m.

Lengths for Curves on Crest Vertical Curves
meters
(or feet)

V (mph)	18.6	24.8	31	37	43	50	56	62	68	74	81
V (km/h)	30	40	50	60	70	80	90	100	110	120	130
% $\triangle$											
1									35 (115)	105 (394)	175 (574)
1.5							50 (164)	110 (361)	170 (558)	240 (787)	
2					8 (46)	58 (190)	118 (387)	178 (584)			
2.5				8 (46)	48 (157)	98 (321)	158 (518)				
3				35 (115)	75 (246)	125 (410)					
3.5			14 (46)	54 (177)	94 (308)						
4			29 (95)	69 (226)							
4.5		10 (33)	40 (131)	80 (262)							
5		19 (62)	49 (161)								
5.5		26 (82)	56 (184)								
6		33 (108)	63 (207)								
6.5		38 (124)									
7	2 (6.5)	42 (138)									
7.5	6 (20)	46 (151)									
8	9 (30)	49 (161)									
8.5	12 (39)										
9	15 (49)										
9.5	17 (56)										
10	20 (66)										
10.5	21 (60)										
11	23 (75)										
11.5	25 (82)										
12	26 (85)										
12.5	28 (92)										
13	29 (95)										

Figure 7.12   Lengths for curves on crest vertical curves, meters (or feet).

Lengths for Curves on Sag Vertical Curves
meters
(or feet)

V (mph)	31	37	43	50	56	62	68	74	81
V (km/h)	50	60	70	80	90	100	110	120	130
% Δ									
1									
1.5									
2									
2.5				29 (95)	47 (154)	65 (213)	83 (272)	104 (341)	125 (410)
3				68 (223)	93 (305)	118 (387)	143 (469)	172 (564)	201 (659)
3.5			70 (167)	95 (311)	125 (410)	155 (508)	185 (607)	220 (721)	255 (836)
4		65 (213)	88 (288)	116 (380)	150 (492)	183 (600)	217 (712)		
4.5		77 (253)	101 (331)						
5	60 (197)								

Figure 7.13   Lengths for curves on sag vertical curves, meters (or feet).

Drainage release points

An important consideration in designing city street profiles is providing drainage release points. If the storm water inlets do not function for some reason, or if a storm exceeds design capacity, the storm water will rise until it reaches an elevation at which it can flow from one drainage basin to the next basin. That elevation and its horizontal location is the drainage release point. First, an acceptable depth of flooding must be determined. The release points must be no higher above the low point in the drainage basin than the acceptable depth of flooding (Fig. 7.16).

When a subdivision is to be designed in a very flat area, the natural fall of the land may not allow for a 0.003 slope. Here, slopes must be created with a zigzag design in the profiles. The profile should be designed with a grade of −0.003 for a distance and then with a grade of +0.003. The length of the positive slope must be shorter than the length of the negative slope, so that there will be a negative slope between successive high points (Fig. 7.16) of no less than 0.0005. This is called the *piezometric slope* and is necessary for the drainage to flow.

Care should be taken to determine the drainage release point at street intersections. Where streets cross, intersecting curbs are connected with short-radii, 6 to 15 m (20- to 50-ft) curves. These are called *curb returns*. The radius to use will be dictated by the approving agency and will depend on the character of the traffic. These connections need to be examined for drainage. There should be enough fall from one end of return (ER) to the other to provide a minimum slope of 0.004 around the curb return. If this slope cannot be provided, one of the street profiles may have to be redesigned. An alternative is to increase the cross slope of one of the streets at the end of the curb

PROFILE CALCULATIONS

Station	Slope	VC	Elev. on tangent	Vert. offset	Elev.	Station	Slope	VC	Elev. on tangent	Vert. offset	Elev.

Figure 7.14 Profile calculations form.

return enough to provide a clear low point of drainage. It is desirable to locate the low point at the end of return (ER). Placing catch basins within curb returns should be avoided.

Abrupt changes in the profiles of the curbs are undesirable. Where the difference in street grades exceeds 5 percent or the elevations at the ends of the returns differ by more than 0.15 m (0.5 ft), profiles of the curb returns should be drawn and curves used where necessary for smooth transitions. Curves can be calculated or a simple profile can be

PROFILE CALCULATIONS

Station	Slope	VC	Elev. on tangent	Vert. offset	Elev.	Station	Slope	VC	Elev. on tangent	Vert. offset	Elev.
49 + 00			223.03		223.03						
50 + 00	−0.034 = S				219.63						
51 + 00					216.23						
51 + 25.80	New ₵	=			215.35						
51 + 73.63	EX ₵										
52 + 00					214.45						
53 + 00					211.05						
53 + 50		BVC	209.35	0	209.35						
54 + 00		200-ft VC	207.65	0.09	207.56						
54 + 50	PI		205.95	0.35	205.60						
55 + 00			203.55	0.09	203.46						
55 + 50	−0.048	EVC	201.15	0	201.15						
56 + 00					198.75						
57 + 00					196.35						

Figure 7.15 Profile calculations form filled out.

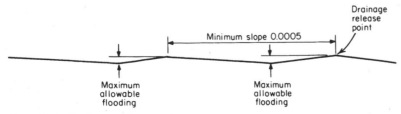

Figure 7.16 Drainage release point.

drawn. Elevations at the midpoint of the return and at quarter points can then be measured from the profile and marked on the plan view (Fig. 7.17).

Roadway Cross-Section Design

The approving agency usually dictates the geometric and structural cross sections based on traffic and soil foundation factors. In most private developments, streets will eventually be dedicated to public use and will become the responsibility of the public agency to maintain.

Geometric cross sections

Geometric cross sections are used beginning in the planning stages of projects. The widths and dimensions of the various elements of the roadway are shown. A standard width for traffic lanes is 3.6 m (12 ft). However, on projects that are renovations of existing facilities, the design width may be as narrow as 3 m (10 ft). Also shown on the geometric cross section may be right-of-way widths, widths of shoulders, bicycle lanes, medians, curbs and gutters, sidewalks, sound walls, median barriers, and planting strips (see Fig. 7.18).

Cross slopes are shown on geometric cross sections. Cross slopes range from 2 to 4 percent on traveled ways and up to 5 percent on shoulders. Where roadways are in horizontal curves, superelevations are designed that provide cross slopes of up to 12 percent to provide better control of the vehicle in overcoming centrifugal force.

Structural cross sections

The types of materials to be used for the construction of roadways, their thicknesses, and other specific information including geometrics are shown on the structural cross section (see Fig. 7.19). The choice of

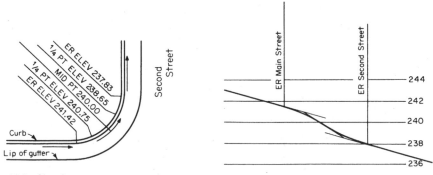

Figure 7.17   Curb return.

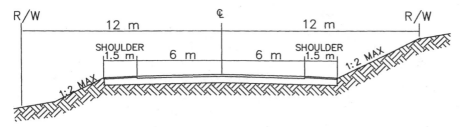

Figure 7.18 Geometric cross section.

material is based on expected traffic loads and the characteristics of the supporting soil. A soils engineer should make recommendations. The types of curb required and the type of sidewalk are shown.

A number of materials are used to comprise the structural section. The top layer is usually asphaltic concrete (AC) or portland cement concrete (PCC). Beneath the top layer there may be a layer of permeable material to collect any moisture that has seeped through the top layer. In this layer there will also be an edge drain, which will collect and carry the moisture away. Next is an aggregate base (AB) layer and then possibly an aggregate subbase (AS) layer. Cement-treated base (CTB) or lime-treated base (LTB) are also recommended under certain conditions. If groundwater is likely to be a problem, a layer of permeable material wrapped in a filter fabric may be required to ensure that the groundwater does not invade the base material and cause a loss of integrity. French drains (Fig. 6.1) may be recommended at the edges of paved areas to draw down groundwater so that it does not reach the subbase. A soils engineer would recommend depth and locations for the french drains. The earth foundation is constructed with 95 percent compaction.

Curbs and gutters

There are six basic types of curbs (Fig. 7.20).

1. *Curb and gutter.* Typically the gutter is 0.45 to 0.60 m (1.5 to 2.0 ft) wide (Fig. 7.20a) and is sloped toward the curb to carry drainage along the edge of the roadway to a drainage inlet.

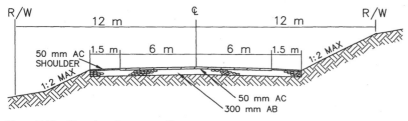

Figure 7.19 Structural cross section.

2. *Vertical curb with spill gutter.*   The gutter is sloped away from the curb and is used along medians and in other places where the roadway is sloped away from the curb (Fig. 7.20b).
3. *Rolled or mountable curb*
    a. When roadway improvements are built before improvements to the lots are planned, a rolled curb may be used. The advantage of the rolled curb is that it can be driven over. This way, when improvements are made to the lots, the driveways can be located anywhere along the property without requiring that the curb be removed and replaced by a driveway cut. The disadvantage of using rolled curbs is that driving over them can adversely affect the front-end alignment on cars, and they provide less protection for pedestrians (Fig. 7.20c).
    b. On highways, a rolled curb may be used to transport drainage while allowing vehicles to move onto the shoulder and beyond in an emergency.
4. *Vertical curb.*   This curb extends below as well as above the pavement and includes no gutter. It is used around medians, islands, and planters (Fig. 7.20d).
5. *Stick-on (extruded) curb.*   This concrete curb can be used in parking lot design. It is attached to the paved surface with an epoxy (Fig. 7.20e).
6. *Asphalt berm or dike.*   This asphaltic concrete (AC) curb is usually used on roadways in rural areas to direct drainage (Fig. 7.20f).

Superelevations

On straight sections of streets and roads, the cross slope ranges from 1 to 4 percent away from the centerline so that the water will drain off. This is called a *crown.* When the horizontal alignment of roads is curved and speeds greater than 60 km/h (30 mph) are expected, centrifugal force becomes a factor affecting a safe, comfortable ride. To overcome centrifugal force, the cross slope is tilted up on the outside of the curve. This tilting is called *superelevation* (Fig. 7.21).

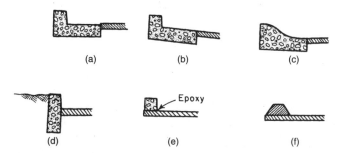

Figure 7.20   Typical curbs: (a) curb and gutter; (b) curb with spill gutter; (c) rolled curb and gutter; (d) vertical curb; (e) stick-on curb; (f) asphalt berm.

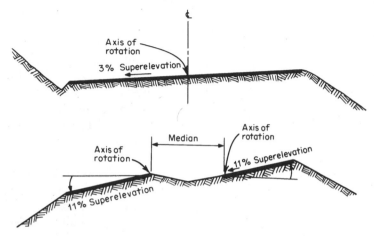

Figure 7.21   Superelevations.

The rate of the cross slope is determined by the design speed, the degree of curve, and the friction of the tires on the pavement. The approving agency will have established standards and methods for determining the superelevation slopes to be used and for designing superelevation transitions.

On private driveways built on the side of a hill, the road may be sloped toward the cut side to facilitate drainage regardless of horizontal curves. This is because if the driveway drainage is allowed to flow over the embankment, erosion will occur. Of course, whether this can be done safely depends on the expected traffic speed, which in turn is based on the length and steepness of the driveway slope.

Widening Existing Roadways

Often it is necessary to meet a half street and complete it to full width or to widen an existing roadway and construct curbs. When this is the assignment, an examination of the existing roadway is essential. Determine the width and condition of the existing pavement. The following questions must be answered:

1. Should the existing pavement be removed and replaced?

2. Will an asphalt overlay be needed?

3. Should a geotextile be employed to prevent reflective cracking?

4. If the existing traveled way is to remain, should the new section overlay the old one to conform, or should a sharp edge be cut and a butt joint made?

5. Is there a shoulder or temporary paving that will have to be removed?

The approving agency may dictate the answers to these questions, but you should satisfy yourself that they are the right choices. When the assignment is to meet an existing edge of pavement longitudinally, the existing edge of pavement must be surveyed at vertical break points and every 8 m (25 ft). The edge of pavement should then be plotted at an exaggerated scale, such as 1:500 (1 in = 20 ft) horizontally, 1:50 (1 in = 2 ft) vertically. Use of the exaggerated scale makes any deviation from a smooth profile apparent. Profiles should also be plotted where the new edge of the pavement will fall if the cross slope is 1 percent and also if the cross slope is 4 percent.

If the new edge of the pavement is to be 8 m (25 ft) from the old one, the new edge-of-pavement profile will fall between 0.08 and 0.3 m (0.25 and 1.0 ft) below the existing edge-of-pavement profile. An edge of the pavement within the area between these profiles will provide a satisfactory cross slope (Fig. 7.22). If the existing edge of the pavement is very uneven, dips may be filled in with asphalt, or some other design must be devised to provide a smooth ride.

The design of a new curb should be handled in a similar manner. Calculate the elevation difference between the longitudinal joint and the new curb for a 1 percent cross slope and a 4 percent cross slope. If the new cross section is to include a shoulder, the elevation difference should be calculated to include the cross slope of the shoulder. Draw the profiles of these elevations. Design the profile of the new curb to fall between the two profiles. Make the cross slope of the new section match the cross slope of the existing section as much as possible.

Cross sections are helpful in coordinating cross slopes and profiles. If there is to be an asphalt overlay of the existing section, the depth of the overlay must be added to the existing profiles when calculating the new profiles. Be alert to existing facilities. Wherever overlays are used, manhole rims and covers, valve covers, and other facilities in the roadway will have to be adjusted to grade and that work described on the plans and in the specifications. If sidewalks and curbs are being added or relocated, underground utility lines, transformers, or junction boxes may be affected.

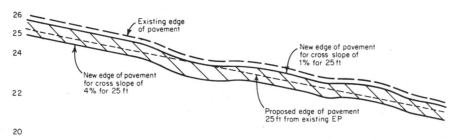

Figure 7.22   Matching the existing edge of pavement longitudinally.

Parking Lots

The layout of parking lots is usually designed by architects or planners and is based on requirements for driving lanes and parking spaces of certain dimensions. Also a certain number of parking spaces per square meter of commercial or industrial space will be required by the approving agency. It is the engineer's task to design the slopes and elevations of the parking lot so that the drainage flows away from structures and does not present a hazard for people using the parking lot. The engineer is also responsible to verify, with the use of truck turns, that the parking lot provides adequate routes for fire trucks, delivery trucks, and garbage collection trucks.

The amount and direction of fall across the lot will influence, if not dictate, how the lot should be drained. Some jurisdictions will not allow parking lots to drain over the sidewalk. This precludes draining the lot over the driveway as well. In such cases, drainage inlets must then be provided at the back of walk.

The elevations and conditions at the boundary influence how to approach the design. Determine if it will be necessary to provide ditches or retaining walls along the property lines. Parking lots can be sloped to drain toward curbs and gutters at the sides or toward area drains in the driving lanes. There may be an increased risk of storm water inlets being blocked by a tire or an accumulation of trash where drainage is at the side, but there are significant advantages.

If the curb and gutter at the side (Fig. 7.23) is used, the drainage in the longitudinal direction can be at a slope of 0.004. This flat slope will

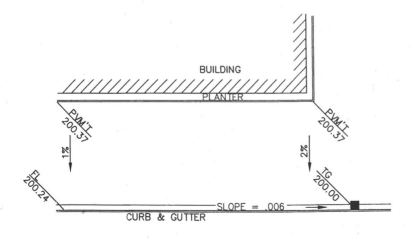

FGI = FLAT GRATE INLET
TG = TOP OF GRATE
PVM'T = PAVEMENT

Figure 7.23 Paved area drained to curb and gutter.

allow a greater distance between inlets, resulting in fewer inlets. Also, less zigzagging of the profile will be required. A concrete curb and gutter is strong and will provide enduring drainage control.

A nearly flat slope can also be accomplished with a valley gutter (Fig. 7.24). It is more expensive to build a valley gutter down the driveway and a curb at the edge of the parking area than simply to build a curb and gutter at the edge. Valley gutters in driveways are more subject to failure as well. However, valley gutters are clearly indicated for alleyways and in multifamily projects where garages are on either side of the driveway and there is little design flexibility.

A popular way to drain parking lots is to divide them into sections and drain toward the center of each section. If drainage reaches of 15 m (50 ft) or less are used, a minimum slope of 1 percent can be used. Slopes must be set in critical directions. The critical direction is often between the inlet and the farthest corner of the area draining into it. The slopes must then be checked between the inlet and all other high and low points in the area of drainage (Fig. 7.25) to verify that the minimum and maximum slope criteria are not exceeded. This method works well, and the slope is not noticeable to those using the lot. If drainage reaches longer than 15 m (50 ft) are chosen to minimize the number of inlets required, the minimum slope should be 2 percent. When 2 percent is used as the minimum slope in the longitudinal direction, the slopes in the short direction are much steeper, and a bowl effect results. Architects and others concerned with aesthetics often object to this bowl-like appearance. Four percent should be the maximum cross slope used in parking areas (Fig. 7.26). This is one of the reasons that a curb and gutter at the side is preferred over the central inlet.

Drainage release points (Fig. 7.16) for parking lots must be examined. The lots should be designed so that if all the inlets do not function, the storm water will still flood to no more than 0.3 m (1 ft) deep before spilling over into the adjacent area.

Handicap Parking and Ramps

The Americans with Disabilities Act (ADA) of 1990 spelled out criteria for accommodating citizens in wheelchairs, those who are sight or

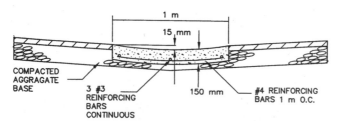

Figure 7.24 Valley gutter.

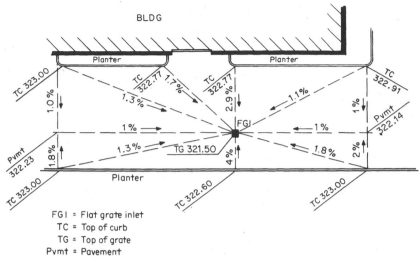

FGI = Flat grate inlet
TC = Top of curb
TG = Top of grate
Pvmt = Pavement

Figure 7.25 Paved area drained to center. Slopes checked.

hearing impaired, and others with limited abilities. Architects and planners are usually responsible for compliance with most of the specifications for buildings and access but civil engineering site designers need to be aware of some of the criteria. The information given here is very general. The jurisdiction in which you are working may have different criteria to be met. Further, there are historic preservation sites, medical facilities, and other situations where requirements are different. For unusual sites, check with the latest information and amendments to the ADA. Check with local jurisdictions and references at the end of this chapter.

Figure 7.26 Car on 7% cross slope.

Parking should be provided at the rate on this table:

Handicapped Parking Spaces ADA Requires

Total spaces	Handicapped parking spaces
1 to 25	1
26 to 50	2
51 to 75	3
76 to 100	4
101 to 150	5
151 to 200	6
201 to 300	7
301 to 400	8
401 to 500	9
501 to 1000	2 percent of total
1001 and over	20 spaces plus 1 for each 100 over 1000

The parking spaces also must meet certain criteria. There must be an aisle next to the parking spaces 1.5 m (5 ft) wide. The parking spaces must be marked with the international symbol for handicapped parking and must be signed with the symbol and the words "PARKING ONLY." At least one handicapped parking space on every parking lot must be signed "VAN ACCESSIBLE." The size of the aisle next to the van-accessible space must be 2.44 m (12 ft) wide minimum which can be shared with a van on the other side of it. There must be one van space for every eight handicapped parking spaces. The slopes of the parking spaces and aisle and the cross slopes on the sidewalks along the accessible route must not exceed 2 percent.

The ADA requires that there be accessibility to entrances not only from the parking lots but from public transportation and public streets as well. This means that sidewalks must be at least 915 mm (36 in) wide continuously and nowhere be less than 815 mm (32 in) wide along the accessible route. In a situation where it is likely that two wheelchairs might pass, the width must be 1520 mm (60 in). The slope shall not exceed 1:20 (5 percent) along access routes. If 5 percent slope is not possible due to site conditions, a slope of up to 1:12 (8.33 percent) can be used but the criteria established for a *ramp* must be adhered to.

Ramps must have a clear width of 915 mm (36 in) and must have hand railings and curbs a minimum of 50 mm (2 in) high. They may rise at a rate of 1:12 for a maximum vertical distance not to exceed 760 mm (30 in). That allows a maximum distance of 9 m (30 ft). For every rise of 760 mm (30 in), there must be a landing of 1525 mm (60 in). Where the ramp changes direction, the landing must be a minimum of 1525 mm (60 in) square. Most public agencies have standard plans of ramps that allow wheelchairs to move from streets to public sidewalks.

Summary

The first task in designing roadways is establishing a centerline or reference line. All other appurtenances will be tied to that reference line.

In most cases, the technical aspects of the design simply require an understanding of the geometry of horizontal curves and an ability to calculate some simple equations for vertical curves. An understanding of horizontal and vertical passing and stopping sight distances is also needed. The geometric and structural cross sections provide the design for the pavement surface and are based on foundation soils and the expected quality and quantity of traffic.

Layout of parking lots is usually performed by a planner, but the engineer should prepare the vertical design to provide for drainage. Parking lots and other aspects of site design require an understanding of methods for providing access to people with disabilities.

Problems

1. What is the major difference in design considerations between rural roadways and city streets?

2. What factors influence the horizontal location of an urban highway?

3. How are the courses of the reference line described?

4. Where are stationing equations needed?

5. What information is needed to describe a circular curve?

6. How do you determine the length of curb return? Why is that length needed?

7. What information should be shown on the profile?

8. What factor determines the maximum allowable slope on highways?

9. What determines the minimum allowable slope on city streets?

10. What length vertical curve should be used for adequate stopping sight distance for vehicles traveling 40 km/h (25 mph)? The grades are +6 percent into the curve and –2 percent coming out of the vertical curve. Use Fig. 7.12.

11. When the profile slope goes from a positive slope to a steeper positive slope, do you have a crest or a sag curve?

12. Why is the critical point on a sag curve important?

13. Calculate the location and elevation of the critical point on the following curve:

Slope is –3.5 percent going in.

Slope is +5 percent going out.

Use 350 m vertical curve.

The PI is at STA 10+22 with an elevation of 200.

14. What are truck turns, and why are they important?

15. What is the difference between geometric and structural cross sections?

16. What is a superelevation?

17. What is the maximum cross slope to use in a parking lot?

18. What is the maximum sidewalk slope on the handicapped access route?

Further Reading

American Association of State Highway and Transportation Officials, *A Policy on Geometric Design of Highways and Streets,* Washington, D.C. 1984.

State of California, Department of Transportation, *Highway Design Manual Metric,* Sacramento, California, 1995.

State of California, Department of Transportation, *Truck Paths on Short Radius Turns,* Sacramento, California, 1972.

U.S. Architectural & Transportation Barriers Compliance Board, *Americans with Disabilities Act (ADA), Accessibility Guidelines for Buildings and Facilities,* http://codi.buffalo.edu/text//legislation/.adaag/.toc, Federal Register, Washington, D.C. 20036-3894, 1991.

8

Sanitary Sewers

Collection and treatment of sewage is the most critical element in the development of any site where people will be spending time. Without a plan for safe disposal of sewage, a site cannot be developed. In most cases, an existing sewer system with available capacity can be connected to with a gravity-flow sewer network. When a site is too low to allow gravity flow or when there are no public systems in the area, the problem becomes more complicated. The design of a gravity-flow sewer network is described in this chapter. The use of force mains, vacuum systems, and septic systems are discussed briefly.

Sources and Quantities

Determining the quantity of sewage is of primary importance when designing a sewerage network. The sanitary sewer district or agency which will treat and dispose of the effluent must be consulted. The agency should have a master sanitary sewer plan based on expected future growth as well as on existing needs. The quantity of sewage per unit and the size of conduits to use within your project may be dictated by the agency. The master plan may require installation of a conduit that is larger than would be necessary to accommodate your project alone. When this is the case, compensation may be paid or credited to the builder for the extent of upgrading the sewer. If the responsible jurisdiction does not dictate quantity, you must determine what quantity to use in designing the system.

Sewage Production

The use of the site, whether residential, commercial, or industrial, influences expected flows. Sewage flows are also related directly to water consumption. About 60 to 80 percent of the per capita consumption of water will become sewage. For this reason, begin by looking at water consumption when other sources of flow-rate information are not available. As illustrated in Table 8.1, consumption rates vary considerably from area to area. Notice that water consumption ranges from 50 gpd/cap (50 gallons per day per capita) in Little Rock, Arkansas, to 410 gpd/cap in Las Vegas, Nevada. In Las Vegas, 51 percent of the water consumed reaches the sewer system. In contrast, in Little Rock, 100 percent of the water consumed reaches the sanitary sewer.

The differences in amounts of water used are affected by such things as climate, distribution of land uses, cost of water, cultural attitudes, and availability of nonpublic sources of water. Table 10.1 lists water consumption based on production volumes for various industries. Table 10.2 lists estimated water consumption for various types of nonindustrial establishments. To estimate sewage flow rates on a particular project, determine the percentage of water consumed that does not reach the sanitary sewer lines. If the project is a park, for instance, a low percentage of the water consumed will reach the sanitary sewer system. In a meat processing plant, on the other hand, a high percentage of the water consumed will reach the sanitary system. The best figures to use in the case of an industrial site are ones supplied by the client. Accurate water consumption figures and an estimate of the percentage expected to reach the sanitary system should be available.

The fixture-unit method

The use of the fixture-unit method for estimating flows from hotels, apartments, hospitals, schools, and office buildings is often indicated. The United States of America Standards Institute National Plumbing Code, USASI A40.8-1955, defines a fixture unit as *a quantity in terms of which the load producing effects on the plumbing system of different kinds of plumbing fixtures are expressed on some arbitrarily chosen scale.* The quantity is approximately 1 cfm (7.5 gpm). Table 8.2 lists some examples of fixture-unit values for various facilities.

The peaking factor

The sewer flow rate varies during the day. Therefore, sewer sizes are not designed for the average flow, but for *peak flows.* The peak flow is the highest instantaneous rate of flow that occurs during a day.

Figure 8.1 is a graph showing hourly variation of sewage flow. The peak flow occurs between 1 and 2 p.m. A *peaking factor* is a multiplier applied to the average flow to yield the largest amount of flow expected. For example, if the average flow expected throughout the day is 400 gpd/cap (gallons per day per capita), the average hourly rate is 16.6 gph/cap. The peak flow may occur between 1 and 2 p.m. and may be 800 gpd/cap or 33.2 gph/cap. Here the peaking factor is 2: the peak flow is two times the average flow. The conduits must be designed to accommodate the peak flow if the system is to function adequately. Depending on the number of sources contributing to the sewer, the peaking factor will vary between 1.3 and 2. The agency may dictate a peaking factor.

Infiltration

On large projects, infiltration of groundwater into the sanitary sewers may be a factor in sizing the pipe. Check with the governing agency as to what infiltration rate should be used. Infiltration rates used are generally in the range of 250 to 500 gpd per inch of diameter per mile of pipe.

The Sewer Network

The sanitary sewer network is usually the first of the utilities to be designed. It is normally a gravity-flow system. That is, the sewage is transported from the site to the treatment facility by the force of gravity—no pumping is necessary. Ordinarily the only available outfalls are previously existing sanitary sewer networks, so there is little flexibility available to the designer.

The storm drain network will also be a gravity-flow system, but there are usually more outfalls available, and alternatives other than connections to existing networks can be used. Another factor is the depth of the sanitary sewer. There is less flexibility in the vertical location of sanitary sewers than of storm drains. Design of the sewer network requires accommodation of many factors: the physical conditions of the existing and proposed site, the physical laws of hydraulics, construction technology, cost considerations, and criteria established by the sanitation agency. The agency may dictate criteria for horizontal and depth locations, velocity of flow, and minimum and maximum slopes, minimum pipe sizes, types, and classes. The agency should have a master sanitary sewer plan that will accommodate sanitary sewer needs on an areawide basis for the future as well as for the present.

Horizontal location of the main

A project master plan should be prepared that includes all existing and proposed underground utilities. The scale should be selected to show the entire site and any off-site connections. A copy of the grading plan or tentative map should serve well as a base map. The project master plan is a working drawing. The existing utilities and proposed surface improvements can be plotted on the computer, but the sanitary sewer and storm drain systems should not. They should be drawn by hand on the master plan. It should be drawn on a hard copy rather than in a computer as it is necessary to see the whole site clearly at once. All existing utilities must be shown and their locations dimensioned horizontally and vertically and available capacities shown.

TABLE 8.1 Some Typical Design Flows

City	Year and source of data	Average rate of water consumption, gpd/cap	Population served, 1000s	Average sewage flow, gpd/cap	Sewer design basis, gpd/cap	Remarks
Baltimore, Md.	—	160	1300	100	135 × factor	Factor 4 to 2.
Berkeley, Calif.	—	76	113	60	92	
Boston, Mass.	—	145	801	140	150	Flowing half full.
Cleveland, Ohio	1946 (6)	—	—	100	—	
Cranston, R.I.	1943 (6)	—	—	119	167	
Des Moines, Iowa	1949 (6)	—	—	100	200	
Grand Rapids, Mich.	—	178	200	190	200	
Greenville County, S.C.	1959	110	200	150	300	Service area includes city of Greenville. Sewers 24-in and less designed to flow ⅔ full at 300 gpd/cap; sewers larger than 24 in designed to have 1-ft freeboard.
Hagerstown, Md.	—	100	38	100	250	
Jefferson County, Ala.	—	102	500	100	300	
Johnson County, Kans.	1958	—				
Mission Township main sewer dist.		70	70	60	1350	Most houses have basements with exterior foundation drains.
Indian Creek main sewer dist.		70	30	60	675	Most houses have basements with interior foundation drains.
Kansas City, Mo.	1958	—	500	60	675	For trunks and interceptors.
					1350	For laterals and submains. Many houses have basements and exterior foundation drains.
Lancaster County, Nebr.	1962	167	148	92	400	Serves city of Lincoln.
Las Vegas, Nev.	—	410	45	209	250	
Lincoln, Nebr., lateral dists.	1964	—	—	60	—	For lateral sewers max flow by formula: peak flow $= 5 \times$ avg flow $+$ (pop in 1000s)$^{0.2}$.
Little Rock, Ark.	—	50	100	50	100	
Los Angeles, Calif.	1965	185	2710	85	*	

	Year					Remarks
Los Angeles County sanitation dist.	1964	200	3500	70†	—	
Greater Peoria, Ill.	1960	90	150	75	800	Based on 12 persons per acre for lateral and trunk sewers, respectively.
Madison, Wisc.	1937 (6)	—	—	—	8500	Maximum hourly rate.
Milwaukee, Wisc.	1945 (6)	—	—	125	300	All in 12 h 250 gpd/cap rate.
Memphis, Tenn.	—	125	450	100	—	
Orlando, Fla.	—	150	75	70	100	
Painesville, Ohio	1947 (6)	—	—	125	190	Includes infiltration and roof water.
Rapid City, S. Dak.	—	122	40	121	600	
Rochester, N.Y.	1946 (6)	—	—	—	125	New York State Board of Health Standard.
Santa Monica, Calif.	—	137	75	92	250	
Shreveport, La.	1961	125	165	—	92	Sewer design 150 gpd/cap plus 600 gpd/acre infiltration. Sewers 24 in and less designed to flow ½ full; sewers larger than 24 in designed to have 1-ft freeboard.
St. Joseph, Mo.	1960	—	85	125	450	Main sewers.
Springfield, Mass.	1949 (5)	—	—	—	350	Interceptors.
Toledo, Ohio	1946 (5)	—	—	—	200	
Washington, D.C. suburban sanitary dist.	1946 (5)	—	2 to 3.3 × avg	100	160	150 gpd/cap was used on a special project.
Wyoming, Mich.	1960	150	50	82‡	400	

* The 85 gpd residential multiplied by peak factor.

† Domestic flow only, ranges from 50 to 90 gpd/cap depending on cost of water, type of residence, etc. Domestic plus industrial flow averages 90 gpd.

‡ Calculated actual domestic sewage flow not including infiltration or industrial flow.

NOTE: Gal × 3.785 = ℓ; gpd/acre × 0.00935 = m^3/day/ha; ft × 0.3 = m; in × 2.54 = cm.

SOURCE: American Society of Civil Engineers and the Water Pollution Federation, *Design and Construction of Sanitary and Storm Sewers*, American Society of Civil Engineers, New York, 1979.

TABLE 8.2   Fixture Units per Fixture or Group

Fixture type	Fixture-unit value as load factor
One bathroom group, consisting of tank-operated water closet, lavatory, and bathtub or shower stall	6
Bathtub (with or without overhead shower)*	2
Bidet	3
Combination sink-and-tray	3
Combination sink-and-tray with food disposal unit	4
Dental unit or cuspidor	1
Dental lavatory	1
Drinking fountain	½
Dishwasher, domestic	2
Floor drains	1
Kitchen sink, domestic	2
Kitchen sink, domestic, with food waste grinder	3
Lavatory	1
Lavatory	2
Lavatory, barber, beauty parlor	2
Lavatory, surgeon's	2
Laundry tray (1 or 2 compartments)	2
Shower stall, domestic	2
Showers (group) per head	3
Sinks	
Surgeon's	3
Flushing rim (with valve)	8
Service (trap standard)	3
Service (P trap)	2
Pot, scullery, etc.	4
Urinal, pedestal, syphon jet, blowout	8
Urinal, wall lip	4
Urinal stall, washout	4
Urinal trough (each 2-ft section)	2
Wash sink (circular or multiple) each set of faucets	2
Water closet, tank-operated	4
Water closet, valve-operated	8

* A shower head over a bathtub does not increase the fixture value.

NOTE: For a continuous or semicontinuous flow into a drainage system, such as from a pump, pump ejector, air-conditioning equipment, or similar device, two fixture units shall be allowed for each gpm of flow.

SOURCE: United States of America Standards Institute National Plumbing Code, USASI A40.8-1955.

The utility for which there is the least design flexibility available should be designed first. This is usually the sanitary sewer. The approving jurisdiction may require that the sewer have a particular location, such as 1.5 m (5 ft) north or east of a street centerline. If the horizontal location is not dictated, locate the sanitary main in the street where it will make the least number of crossings with existing utilities and where the lengths of the laterals will be minimized.

All potential outfalls should be identified. A sanitary sewer *outfall* is an existing main or manhole into which the sewer can be discharged. The invert elevations of the outfalls should be labeled on the master

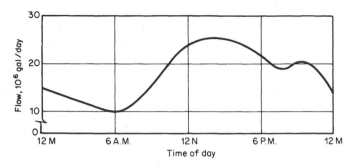

Figure 8.1 Peak flow graph.

plan. The *invert elevation* is the elevation at the inside bottom of the pipe. Using this information and approximate slopes for the pipes, locations of potential crossing conflicts with other utilities can be identified and alternate solutions worked out before much work is done.

Manholes must be provided where sewer mains connect and where there is a change of direction in the main. Some jurisdictions allow horizontal curves for directional changes—some do not. Whatever criteria are dictated must be adhered to. Curves may be allowed if manholes are provided at the beginning and end of the curves or at intermediate locations within the curves. When final construction plans are drawn, relevant curve data should be calculated and shown on the plans. Calculation of curve data is described in Chap. 7. Minimum curve radii or allowable deflections are provided by pipe manufacturers and must be adhered to.

Laterals

Design of plumbing within buildings is ordinarily the responsibility of the architect or mechanical engineer. A connection from the building to the sewer mains should be provided by the civil engineer designing the site improvements. This connection is called a *lateral*.

For commercial and industrial sites, the responsibility for the design of the lateral usually begins at a point 1.5 m (5 ft) from the outside of the building. When the project is to provide the sewer main in the streets for residential lots, the laterals may extend only from the main to the property lines. A sewer cleanout (Fig. 8.2) may be required at the property line. It is desirable to locate cleanout in landscaped areas rather than in driveways. When the cleanout is to be located in a street or driveway, a frame and cover capable of withstanding traffic must be provided. The additional expense of providing cleanout frames and covers for traffic areas is significant on projects where multiple buildings will be served.

When the location and orientation of the buildings and driveways on residential lots have not been determined, the center of the lot is usu-

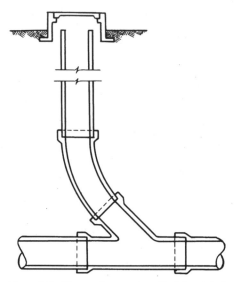

Figure 8.2   Sewer clean out or flushing inlet.

ally a good choice for the lateral. This way, the driveway should fall on one or the other side of the lateral, and the cleanout can be in the lawn or landscaped area.

Backflow prevention or check valves should be provided in laterals where the next upstream manhole on the sewer main is at an elevation above the floor of the building being served (Fig. 8.3). The reason for this is that if the sewer main downstream of the lateral fails to function, the sewage will back up and overflow at the first available opening. Any openings in the system (bathtubs, toilets, and sinks) with openings below the top of the next upstream manhole will provide a relief opening for overflow unless they are protected by a check valve. The check valve should be located where the building plumbing dis-

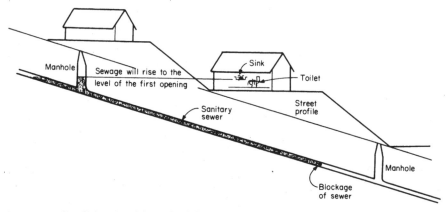

Figure 8.3   Condition requiring a backflow-prevention device.

charges into the lateral. If the top of the next upstream manhole is below the plumbing facilities within the house, the overflow will occur at the manhole.

The lateral should be connected to the main with a wye and a bend. The sewer main must be deep enough to allow the lateral to reach from below the building foundation to the sewer main with a slope of at least 2 percent and to allow for the wye and bend (Fig. 8.4). Sewer laterals can be built through structures, but this additional expense should be avoided whenever possible.

Designing the profile

When a tentative horizontal layout is complete, a profile for the conduit can be designed. On profile paper or in the computer, plot to scale the existing and proposed ground line. It is best to plot the whole length of the profile to use for your design. Plot all crossings of existing utilities and other underground obstacles. Elevations at the top or bottom of crossing utilities should be calculated and marked on the plan. In most cases, the design should be done graphically by hand, not in the computer. A 3 m (9.8-ft) roll showing the whole length of the profile allows you to see what you are dealing with much better than what can be seen on a 500 mm (20-in) screen. This is particularly true where you have a complicated route with a number of utility crossings or other criteria to be met.

If you are working in an area that is densely developed, there might be little or no flexibility as to where the profile will go without impacting existing lines. You may have to thread the pipes through a wall of existing pipes at street crossings. When that is the case, it is those areas that must be designed first since they have the least flexibility. If the new system is to be in a new area where there is no significant criteria that will be a problem, start at the outfall of the new sewer and project the profile upstream. The sewer main must be deep enough to allow for criteria relative to the laterals; 1.5 m (5 ft) is usually the minimum depth.

Where other criteria do not take precedence, the sewer slope should be close to that of the finished grade. The sewer should be kept at a depth

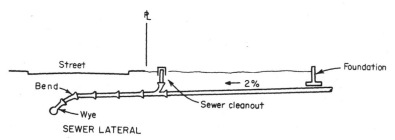

Figure 8.4   Section showing lateral.

as shallow as possible while accommodating the necessary slope on the laterals, space for storm drains and other utility lines and possible protection from freezing. Manholes should be shown at the locations required for the horizontal layout, at grade changes, and at 140 m (450-ft) intervals—more frequently, if required by the approving agency. Keep in mind that a minimum velocity will be required to provide adequate scouring action. The minimum velocity may be based on a stated criteria other than the pipe flowing full. The size of the pipe and the slope selected may provide sufficient velocity when the pipe is flowing at capacity but not during periods of average or less than average flow rate.

When designing sewer profiles, it is important to provide for the fact that nonuniform flow will develop where the slope changes. Where the next downstream slope is steeper, a *draw-down curve* may develop in the profile of the fluid surface. A draw-down curve is the tapering of the longitudinal section. At the steeper slope, the velocity will be greater, so the required area of cross section, and therefore depth, will be less according to the continuity equation (which follows).

When the slope changes from a negative slope to a flatter negative slope, the downstream section will be deeper and a *back-water curve* will develop. This is a gradual increase in the longitudinal cross section. Ordinarily, this nonuniform flow is not a problem. However, if the design calls for a mild slope downstream of a steep slope, the shallow (supercritical) flow may become a deep (subcritical) flow and a hydraulic jump may develop. Sophisticated hydraulics calculations must be performed to understand the impact of the hydraulic jump. One solution to the problem is the installation of manholes to create a gradual change of slope. Manholes with outside drops (Fig. 8.5) can also be installed.

If vertical curves are allowed, use them in the profile where their use will save installation of one or more manholes. When curves are used, they must be calculated as described in Chap. 7. The degree of curvature is limited by the amount of deflection allowed at each section of pipe. The manufacturer should provide information on how much deflection can be made without causing leaks.

When a graphic representation of the profile is complete, the information can be entered into the computer and the pipe sizes, slopes, and inverts can be calculated.

Hydraulics

The science of the mechanics of fluids at rest and in motion is called *hydraulics*. The study of hydraulics is an intellectual and complicated endeavor. Fortunately, however, understanding the use of two hydraulics equations is all that is needed to design the simple gravity-flow sewer networks found in most land development projects. These two equations, the continuity equation and Manning's equation, are discussed in this section. The design of major sewerage trunk lines and

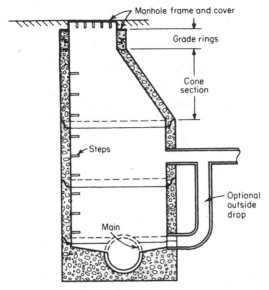

Figure 8.5   Typical manhole.

pressure systems requires more sophisticated techniques that will not be discussed here.

The continuity equation

The most basic hydraulics equation is the *continuity equation*. Stated simply, it says that the quantity (Q) of fluid passing a particular point is the result of its velocity (V) and the cross-sectional area (A) of the flow. The continuity equation is

$$Q = VA \qquad (8.1)$$

where Q = quantity (m³/s)
V = velocity (m/s)
A = cross-sectional area (m²)

The quantity Q is often referred to as the flow rate or capacity. The continuity equation is used to determine the size and slope of the pipe used in a sewer network. The velocity of the flow is a function of the slope and will be discussed under Manning's equation. The size of the pipe is designated by its diameter. Using the required area, the diameter of the conduit can be calculated using the equation for the area of a circle.

$$A = \pi\, r^2$$
$$= \pi \left(\frac{D}{2}\right)^2 \qquad (8.2)$$

Algebraic manipulation yields the value for the diameter (D) as a function of the area.

$$D = \left(\frac{4A}{\pi}\right)^{1/2} \tag{8.3}$$

To determine the diameter, a value for the required area is needed. The continuity equation can be manipulated algebraically to yield the area of the pipe required for a particular velocity.

$$A_R = \frac{Q}{V} \tag{8.4}$$

where A_R = cross-sectional area required (m)
$\quad\quad Q$ = quantity (m³/s)
$\quad\quad V$ = velocity (m/s)

The maximum capacity of a pipe is reached when the pipe is filled to a depth of 0.8 diameter. Beyond this point, the increase in friction reduces the velocity, and thus the capacity. A conservative approximation of the capacity can be made by assuming that the pipe is flowing full.

Example 8.1   Determine the size of pipe needed to transport 0.21 m³/s (7 cfs) of sewage at a minimum velocity of 0.60 m/s (2 fps). Assume the pipe is flowing full.

solution   The quantity (Q) is 0.21 m³/s (7 cfs). The velocity (V) is 0.60 m/s (2 fps).

1. Using Eq. 8.4,

$$A_R = \frac{Q}{V}$$

$$= \frac{0.21 \text{ m}^3/\text{s}}{0.61 \text{ m/s}}$$

$$= 0.34 \text{ m}^2$$

2. Using Eq. 8.3,

$$D = \left(\frac{4A}{\pi}\right)^{1/2}$$

$$= \left(\frac{4 \times 0.34 \text{ m}}{\pi}\right)^{1/2}$$

$$= 0.66 \text{ m}$$

$$= 660 \text{ mm}$$

Pipes are manufactured in various standard diameters. Table 8.3 shows commonly available sizes. When the required diameter for a particular flow falls between standard sizes, the larger size must be used. For the system in Example 8.1, for instance, calculation yielded a pipe diameter of 660 mm (26.22 in). However, only 610 mm (24-in) diameter pipes and 685 mm (27-in) diameter pipes are commonly available. Since 685 mm (26.22 in) falls between the two sizes, the larger 685 mm (27-in) pipe must be used.

This only tells us what pipe will have the desired capacity at a velocity of 0.61 m/s. The problem with this approach for selecting pipe sizes is that it does not provide information about whether the slope we would like to use will meet requirements, nor does it provide the flexibility of using a velocity greater than minimum so that smaller pipes can be used. Manning's equation provides needed relationships about velocity and slope.

Manning's equation

The relationships among the various factors affecting the velocity of flow in pipes are expressed in Manning's equation. Though there are other formulas relating the factors of flow, Manning's is the most commonly used. Manning's equation is given in Eq. 8.5.

$$V = \frac{R_H^{2/3}\, S^{1/2}}{n} \tag{8.5}$$

where V = velocity (m/s)
 n = coefficient of roughness
 R_H = hydraulic radius (m)
 S = slope (m/m)

(In the English system, there is a multiplication factor of 1.49 which accounts for dimensional differences. When metric dimensions are

TABLE 8.3 Pipe Sizes

Pipe size inside diameter, in	Nominal metric pipe size, mm	Pipe size inside diameter, in	Nominal metric pipe size, mm
4	100	33	840
6	150	36	910
8	200	42	1070
10	250	48	1220
12	310	54	1370
15	380	60	1520
18	460	72	1680
21	530	78	1830
24	610	84	2130
27	690	90	2440
30	760	96	—

used, that factor is 1.00.) The n is referred to as Manning's n and is a friction factor for the roughness of the pipe. The value of n to use may be dictated by the responsible jurisdiction. Otherwise, 0.010 can be used for PVC (polyvinyl chloride) and 0.013 can be used for concrete, reinforced concrete, or vitrified clay pipe (CP, RCP, or VCP). Values of n for other materials are given in Table 9.2. The smoother the conduit, the smaller the value of n is. From Manning's equation (Eq. 8.5), we see that the smoother the pipe used, the greater is the velocity, thus capacity, produced. R_H in the equation is the hydraulic radius. It accounts for the effect of friction on the flow. The value of R_H is expressed in Eq. 8.6.

$$R_H = \frac{a}{p} \qquad (8.6)$$

where R_H = hydraulic radius
  a = cross-sectional area of the flow (m²)
  p = wetted perimeter (m)

The wetted perimeter (Fig. 8.6) is the length measured on the cross section that will be wet when the pipe is flowing at the designated capacity. Using the continuity equation and Manning's equation, the quantity can be calculated. For design purposes on simple projects, pipes can be assumed to be flowing full. Assuming that the pipe is flowing full yields a conservative capacity.

Example 8.2   Find the capacity of a 300 mm (12-in) conduit flowing full and installed at a slope of 0.005. Use 0.010 as Manning's n.

solution   We are given $S = 0.005$, $D = 300$ mm (12 in), and $n = 0.010$. The formula for flow rate is $Q = AV$ (Eq. 8.1).

1. Calculate the cross-sectional area of the 300 mm (12-in) diameter pipe.

$$A = \frac{\pi D^2}{4}$$

$$= \frac{3.14 \, (0.305 \text{ m})^2}{4}$$

$$= 0.073025 \text{ m}^2$$

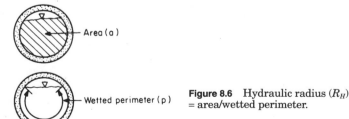

Area (a)

Wetted perimeter (p)

Figure 8.6   Hydraulic radius (R_H) = area/wetted perimeter.

2. Calculate velocity V using Manning's equation.

$$V = \frac{(R_H)^{0.66}\,(S)^{1/2}}{n}$$

We have been given n and S. We must calculate R_H:

$$R_H = \frac{a}{p}$$

As the pipe is flowing full, the area equals the area of the pipe; the wetted perimeter equals the circumference.

$$R_H = \frac{\pi\,D^2/4}{\pi\,D}$$

All the factors cancel except $D/4$.

$$R_H = \frac{D}{4} = \frac{300\text{ mm}}{4}$$

$$R_H = 76\text{ mm}$$

Inserting values into Manning's equation, we get

$$V = \frac{(0.076)^{0.66}\,(0.005)^{1/2}}{0.010}$$

$$= \frac{(0.18)\,(0.71)}{0.010}$$

$$= 1.29\text{ m/s}$$

3. Now all the factors are available. The flow rate can be calculated using Eq. 8.1.

$$Q = AV$$

$$= (0.073\text{ m}^2)\,(1.29\text{ m/s})$$

$$= 0.09\text{ m}^3/\text{s}$$

Calculating the Profile

Begin calculation of elevations and slopes at the outfall. The exact invert elevation of the outfall should have been checked in the field by a survey crew. Where the outfall is to be at an existing conduit rather than an existing manhole, the inverts at the manholes at either end of the existing conduit to be used should be field-checked, a slope between them determined, and an elevation for the proposed outfall calculated. The new connecting line must be equal or smaller in diameter. Even where the connection is made at a manhole, it is imperative that each successive downstream conduit be larger than the last regardless of the size needed to accommodate the needed capacity. Otherwise some particle which passed the upstream conduit might not be able to pass through the smaller conduit and thereby cause an ongoing maintenance problem. Whenever different size pipes are connected, the overts (inside tops of pipes) should be matched in elevation (Fig. 8.7) not the inverts.

Example 8.3 Determine the elevation of the invert for a 460 mm (18-in) conduit connecting to an existing 685 mm (27-in) conduit. The invert elevation of the 680 mm (27-in) sewer is 132.27.

solution Determine the difference in pipe diameters in millimeters and add the difference to the invert of the larger pipe.

$$685 \text{ mm} - 450 \text{ mm} = 235 \text{ mm}$$

$$132.27 + 0.235 \text{ m} = 132.50$$

When calculating the slopes on conduits, care must be taken not to use the difference in stationing as the length of the conduit except where the conduits are straight and parallel with the centerline. The length of the conduit must be determined from the plan view. If the sewer lines are straight where the centerline curves, the conduit will be shorter than the centerline. A conduit will also be shorter when it is concentric with and inside a centerline curve. It will be longer when it is concentric with and outside the centerline curve. If the horizontal location has been established and drawn on the computer, it is a simple matter to label the exact length of the pipe. However, the designer must be aware that the length given by the computer may be to the edge of a graphic manhole. Ordinarily the lengths are from center of manhole to center of manhole so it may be necessary to add length to accommodate that difference. If a computer is not being used at this point, there is another simple way to determine the length of concentric pipe.

Example 8.4 Using Fig. 8.8, calculate the length of the sanitary sewer between centerline STA 5+00 and STA 8+92.70. (For the purpose of this example, it does not matter whether meters or feet are used).

solution The lengths of the curves for the centerline and the sewer are found using the same formula.

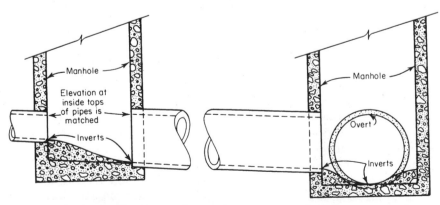

Figure 8.7 Overts (inside tops of pipes), not inverts, are matched.

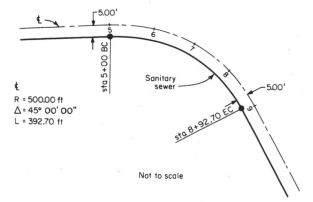

Figure 8.8 Sewer length calculation for Example 8.4.

$$L_{\mathbb{C}} = \frac{\Delta^{\circ}}{360^{\circ}} \, \pi \, 2 \, R \qquad L_{ss} = \frac{\Delta^{\circ}}{360^{\circ}} \, \pi \, 2 \, R_{ss}$$

where K is a constant for each curve with the same angle (Δ°).
Manipulating the equation algebraically gives

$$\frac{L}{R} = \frac{\Delta}{360^{\circ}} \, \pi \, 2 = K$$

$$\frac{L_{ss}}{R_{ss}} = \frac{\Delta^{\circ}}{360^{\circ}} \, \pi \, 2 = K$$

Therefore,

$$\frac{L}{R} = \frac{L_{ss}}{R_{ss}} = K$$

$$= \frac{392.70}{500} = 0.7854 = K$$

$$K = \frac{L_{ss}}{R_{ss}}$$

$$L_{ss} = K \times R_{ss} = 0.7854 \times 495.00 \text{ m} = 388.77 \text{ m}$$

Once the constant for the value of the length divided by the radius for a particular central angle on the centerline curve has been calculated, any other concentric curve with the same central angle (the storm sewer, the curbs, and so forth) can be quickly calculated by multiplying the constant (K) by the radius of the sought-after curve. For Example 8.4, the length of the sanitary sewer between STA 5+00 and STA 8+92.70 is 388.77 m. The slope of the sewer between these stations is based on this distance. Label the actual lengths of the conduits on the profile. The manholes are plotted and labeled at the correct station. When the length of the profile will not scale correctly, label the length "NTS" (not to scale), so it will be apparent that no mistake has been made.

Determine the slope on each section of conduit. Scale the approximate elevations of the inverts at either end from the profile. Subtract the smaller from the larger elevation, and divide that difference by the length (Example 7.3). This yields the slope. Use the approximate slope just determined to work with Manning's formula.

The quantity Q to use for each section of pipe is the quantity at the outfall of each section of pipe. Each lateral connected to the main contributes to the quantity of sewage, but it is impractical to calculate and upgrade the main at each lateral. Therefore, a conduit of sufficient size to accommodate the flow reaching the downstream end is used.

Most jurisdictions require a drop between inverts where sewer mains connect. If that is a requirement, include the required drop wherever necessary. Starting at the outfall invert, calculate the elevation at the upper end of the first section of pipe using the slope and length previously calculated. The elevation of the invert at the lower end of the next upstream section is calculated as the invert elevation at the upper end of the first section plus the differences in pipe diameter or the drop requirement, whichever is larger. This process is continued upstream to the end.

Example 8.5   Determine invert elevations for the sewer pipes shown in Figure 8.9. A drop of 150 mm (0.5 ft) is required for connections at angles greater than 20°. Assume elevations are given in meters above sea level.

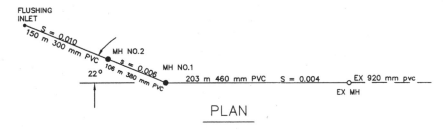

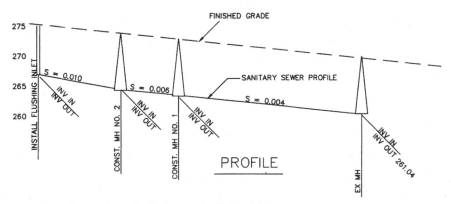

Figure 8.9   Plan and profile for Examples 8.5 and 8.6.

solution

1. The invert elevation at the outfall, a 920 mm (36-in) conduit, is 261.04. The connecting pipe is 460 mm (18 in).

$$920 \text{ mm} - 460 \text{ mm} = 460 \text{ mm}$$

$$261.04 + 460 \text{ mm} = 261.50$$

2. Multiply the length of the next section of pipe times the slope.

$$203 \text{ m} \times 0.004 = 0.81 \text{ m}$$

Add 0.81 to the downstream invert.

$$261.50 + 0.81 \text{ m} = 262.31$$

Thus 262.31 is the invert of the upstream end of the 460 mm (18-in) pipe.

3. The next upstream section of pipe enters MH no. 1 at an angle of 22°. A 0.15 m (0.50-ft) drop is required. The incoming pipe at MH no. 1 has a 380 mm (15-in) diameter.

$$460 \text{ mm} - 380 \text{ mm} = 80 \text{ mm}$$

$$150 \text{ mm} > 80 \text{ mm}$$

The drop requirement exceeds the size difference. Add 0.15 m (0.50 ft) to the downstream invert.

$$262.31 + 150 \text{ mm} = 262.46$$

4. Multiply the length of the next upstream section of pipe times the slope, and add the product to the downstream invert.

$$106.00 \text{ m} \times 0.006 = 0.64 \text{ m}$$

$$262.46 + 640 \text{ mm} = 263.10$$

The upstream invert elevation of the 380 mm (15-in) line is 263.10.

5. The next upstream section of pipe is 300 mm (12 in).

$$385 - 300 = 85 \text{ mm}$$

There is no drop requirement. Add 85 mm (0.25 ft) to the invert of the 380 mm (15-in) pipe at MH no. 2 for the invert of the 300 mm (12-in) pipe.

$$263.10 + 85 \text{ mm} = 263.95$$

6. Multiply the length of the next section by the slope, and add the product to the invert.

$$150 \text{ m} \times 0.010 = 1.5 \text{ m}$$

$$263.95 + 1.5 \text{ m} = 265.45$$

The invert elevation at the flushing inlet is 265.45.

When working in the metric system in the United States, the only elevations available may be in feet rather than in meters above sea level. When this is the case, care must be taken to make the conversions to feet for vertical dimensions before performing calculations.

Example 8.6   The sewerage system in Fig. 8.9 is in a sanitation district with the following criteria:

Minimum slope = 0.004

Minimum pipe size = 200 mm (8 in)

Velocity range = 0.6 m/s minimum, 3 m/s maximum (2 fps min, 10 fps max) when the system is flowing full

Manning's $n = 0.010$

Assume the pipes are flowing full. Are the district's criteria met for the following?

a. The conduit from the outfall to MH no. 1 with a peak flow of 0.23 m³/s (8 cfs)?

b. The conduit from MH no. 1 to MH no. 2 with a peak flow of 0.17 m³/s (6 cfs)?

c. The conduit from MH no. 2 to the flushing inlet with a peak flow of 0.08 m³/s (3 cfs)?

solution

a. The conduit in part a is a 460 mm (18-in) PVC at a slope of 0.004. The continuity equation $Q = VA$ will give us the capacity. First, we must determine the area of the conduit and the velocity of the flow. The area is determined by the equation

$$A_{460} = \pi r^2$$

$$= 3.14\ (230\ \text{mm})^2$$

$$= 3.14 \times 52\ 900\ \text{mm}^2$$

$$= 0.17\ \text{m}^2$$

We can determine the velocity using Manning's equation (Eq. 8.5),

$$V = \frac{R^{2/3}\ S^{1/2}}{n}$$

For this, we need the hydraulic radius R_H found from Eq. 8.6,

$$R_H = \frac{a}{p}$$

For this relationship,

$$p = \pi D$$

$$= 3.14 \times 0.460\ \text{m} = 1.44\ \text{m}$$

Therefore,

$$R_H = \frac{a}{p}$$

$$= \frac{0.17 \text{ m}^2}{1.44 \text{ m}} = 0.118$$

The a is the same as A as calculated before.

$$V = \frac{(0.118)^{2/3} (0.004)^{1/2}}{0.010}$$

$$= \frac{(0.244) (0.063)}{0.010}$$

$$= 1.5 \text{ m/s}$$

The velocity of 1.5 m/s (4.9 fps) is in the acceptable range. Now inserting A and V into the continuity equation yields a capacity of the 460 mm (18-in) PVC at a slope of 0.004 of 0.26 m/s (9.18 cfs). The velocity is within the acceptable range and the capacity is sufficient.

Velocity   3 m/s > 1.5 m/s > 0.6 m/s

Capacity   0.26 m³/s > 0.23 m³/s

b. Using the same procedures, the velocity is calculated to be 1.2 m/s (3.9 fps) for a 380-mm (15-in) PVC at a slope of 0.006. The capacity is 0.15 m³/s (5.2 cfs). This section meets the requirements.

c. For this part of the system, the velocity is 1.7 m³/s (5.5 fps) with a capacity of 0.19 m³/s (6.7 cfs). The capacity required is 0.17 m³/s (6 cfs), so this part of the system meets all the criteria.

To accommodate the quantities shown, the system will not have to be redesigned. It meets the given criteria. If it did not, it would have been necessary to use larger pipes or steeper slopes.

The problem with these examples, of course, is that in the real world you will not be given such convenient information. What you will have is the slope that you designed and the quantity that you need to accommodate. What you have to do is assume a pipe size, plug in and manipulate the formulas, and see what capacity the assumed size will accommodate. If it is too small, you will have to recalculate using a larger size. If it is much bigger than needed, you will want to recalculate to see if a smaller size will work. It is an iterative process.

For each design alternative considered, clearances with other utilities and coverages must be checked. You will develop a sense for what will work after working with sewer systems over a period of time, but the calculations will still be time-consuming. Sophisticated CADD software systems are available that will make the calculations, offer alternatives, and draft the designs. Software that does not require CADD is also available for making hydraulic calculations and offering alternatives. These calculations can also be entered into a spreadsheet program that will then allow the designer to try different pipe sizes and slopes and obtain capacities instantly. Figure 8.10 illustrates the information and formulas entered into a Microsoft Excel spreadsheet, and Fig. 8.11 shows the results when the information from Example 8.6 is input. Different spreadsheet software may require different input. The engineer can custom-design the spreadsheet to the particular requirements of a specific project.

Manholes and Flushing Inlets

Maintenance of the sewer network is necessary, so manholes are installed at regular intervals and at potential trouble spots—where the direction of flow changes either horizontally or vertically and where connections are made. A manhole or flushing inlet (see Fig. 8.2) should be installed at the top end of a sewer line to allow maintenance crews to flush out the line and check for breaks.

The jurisdiction should have standard plans and specifications for manholes and flushing inlets. Manholes are typically 1.2 or 1.5 m (4 or 5 ft) in diameter and are topped by a cone (see Fig. 8.5). The cones provide the transition between the 1.2 or 1.5 m (4- or 5-ft) diameter manhole and the 0.6 m (2-ft) opening at the street. Openings in cones can be concentric but are more often eccentric. With eccentric cones, the ladder of the manhole is constructed on the straight side. When the manhole is located on the street or some other traffic area, the ladder and cone should be turned so that the maintenance person will be facing traffic when entering and leaving the manhole. Manhole covers are round so that they cannot be dropped into the manhole and so that there is less likelihood of their rattling when driven over.

Occasionally, it will be necessary to locate a manhole where the cover will conflict with a curb or some other structure. In this case, if the conflict is small, turning the cone to affect the location of the cover may be helpful. Grading rings may be placed on top of the cone before the frame and cover to extend a deep manhole to the surface. To protect the health and safety of maintenance personnel, conduits should connect no higher than 1 m (or 2.5 ft) above the bottom of the manhole. If this is impractical, an outside drop should be constructed (Fig. 8.5).

The inside bottoms of sanitary manholes should be shaped to ensure smooth flow of the sewage. One way to accomplish this is to set bends of sewer conduits in the concrete (Fig. 8.12). When the concrete has hardened, the top half of the pipe is broken out and connections are made smooth with cement. When several conduits are to enter one manhole, draw a plan view and a cross-sectional view at a large scale, such as 1:20 (1 in = 2 ft), to check that there is sufficient surface to accommodate the conduits.

Example 8.6 Sanitary Sewer System Calculations for Fig. 8.9

Row #	Point of concentration A	Total Q (m³/s) B	Diameter D (meters) C	Circumference Wetted Perimeter D	Pipe area (m²) E	Hydraulic Radius (R_H) F	Slope (m/m) G	Velocity (m/s) H	Length (m) I	Q=VA Capacity (m³/s) J	Invert in K	Invert out L	Top of pipe elevation M	Cover N
12	Flushing inlet	0.30											=L12+.305	274.5-M12
13			.305	=C13*3.14	=(C13/2)²*3.14	=E13/D13	.012	$=(F13)^{.66}(G13)^{.5}/.010$	46	=H13*E13		=L14+.025		
14	MH # 2	0.17									=M14+0.08		=L14+.386	272.5-M14
15			.385	=C15*3.14	=(C15/2)²*3.14	=E15/D15	.006	$=(F15)^{.66}(G15)^{.5}/.010$	32	=H15*E15		=L16+G15*I15		
16	MH # 1	0.08									=M16+0.08		=L16+.460	271.5-M16
17			.460	=C17*3.14	=(C17/2)²*3.14	=E17/D17	.004	$=(F17)^{.66}(G17)^{.5}/.010$	62	=H17*E17		=L18+G17*I17		
18	EX MH										=L18+0.46	261.04	=L18+.460	268.5-M18

Figure 8.10 Spreadsheet showing formulas in cells for Example 8.6.

Example 8.6 Sanitary Sewer System Calculations for Fig. 8.9

Row #	Point of Concentration	Total quantity required (m/s)	Diameter D (meters)	Circumference Wetted Perimeter R	Pipe Area (m)	Hydraulic Radius (R)	Slope (m/m)	Velocity (m/s)	Length (m)	Q=VA Capacity (m/s)	Invert in	Invert out	Top of Pipe Elevation	Cover
	A	B	C	D	E	F	G	H	I	J	K	L	M	N
12	Flushing Inlet											262.65	262.96	11.54
13		0.08	0.305	0.96	0.07	0.08	0.012	2.0	46	0.15				
14	MH # 2										262.10	262.02	262.41	10.60
15		0.17	0.385	1.21	0.12	0.10	0.006	1.7	32	0.19				
16	MH # 1										261.83	261.75	262.21	9.79
17		0.22	0.460	1.44	0.17	0.12	0.004	1.5	62	0.25				
18	EX MH										261.50	261.04	261.96	6.54

Figure 8.11 Spreadsheet showing values in cells for Example 8.6.

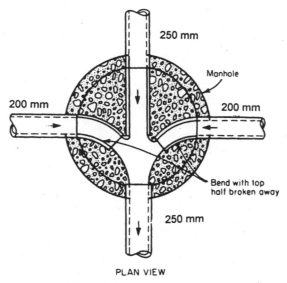

Figure 8.12   Manhole bottom.

Pressure Systems

When existing sanitary sewer outfalls are too high for a gravity-flow system, a pumped system must be used. Pressure systems cannot accommodate laterals, so a gravity-flow system is designed to accommodate the flow while keeping the pipes as shallow as possible. At the lower end of the system, a manhole or wet well is installed. A force main (pressure system) is then designed to carry the sewage from the manhole at the low end of the system to an available outfall. The force mains (FMs) may parallel the gravity system or may be located elsewhere. A pumping station is located at either the upper or the lower end of the force main.

Vacuum Systems

In some situations it may be more economical to construct a vacuum sanitary sewer system. Where new systems would be constructed under existing roadways or where the required depth is excessive, the cost of trenching and of replacing pavement may make it worthwhile to consider using a vacuum system. Because they can be built near the surface, between 1 and 1.5 m deep, the direct cost of trenching would be less. Another advantage is that with the use of shallower trenching, the time needed for construction is less; therefore, there is less disruption of traffic in developed areas. This factor could be an important influence on the political element as well. An article, *Defying Gravity,* in the February 1998 *CE News* magazine describes the use of vacuum systems in more detail.

Septic Systems

In areas where a sanitary sewer system is not available, septic tanks and leach fields may be an alternative for sewage treatment. The size of the septic tank is determined by the expected capacity. The soil must be tested to determine the depth and percolation rate. The length of the leach field is a function of the percolation rate and the needed capacity. The septic tank and leach field will have to be located far enough from structures that they do not affect the structural foundations. Leach fields must be located where they will not contaminate wells, streams, or other water sources.

Summary

Every site where people will spend time must be provided with some kind of sanitation system. Most often the system is connected to a network of sewerage. Usually, the sanitary sewer will have the most limitations and is the deepest of underground facilities, so it is installed first and should be designed first. The sizing of the pipes is based on the continuity equation and Manning's equation. Calculating the profile of the pipes makes use of the techniques used for calculating profiles for roadways.

Problems

1. Why is water consumption important to sewerage systems?

2. What factors influence the quantity of sewage?

3. How much sewage is produced by one fixture-unit?

4. What is a peaking factor?

5. What is the reason for requiring a minimum velocity?

6. Why are backflow-prevention devices necessary?

7. Where should backflow-prevention devices be installed?

8. What is a minimum slope for a sewer lateral?

9. Before designing a sewer main, what information should be plotted on the profile?

10. What is the outfall?

11. Where are manholes placed?

12. What standard size VCP is needed to accommodate 0.6 m³/s (21 cfs) of sewage flowing at 1 m/s (3.3 fps)? What size PVC is required? Assume the pipe

is at a slope of 0.005 and is flowing full. This may be an iterative process. Try one size. If it doesn't work try another.

13. What is the wetted perimeter of a 760 mm (30-in) pipe flowing full?

14. What is the hydraulic radius of a 910 mm (36-in) pipe flowing half-full?

15. Find the velocity of flow in a 460 mm (18-in) pipe flowing full at a slope of 2 percent. Manning's n is 0.010.

16. What is the capacity of the 100 mm (4-in) lateral in problem 8? Manning's n is 0.010.

17. The upstream end of a 760 mm (30-in) pipe has an invert of 134.75. What invert should be used for the downstream end of a connecting 460 mm (18-in) pipe?

18. The sanitary sewer is located 1.5 m (5 ft) south from the center of a street with a 137 m (450-ft) radius curve convex to the north. The invert elevation at the BC STA 102+61 is 708.34. The invert elevation at the EC STA 105+52 is 702.15. What is the slope of the pipe?

Further Reading

American Society of Civil Engineers Staff and Water Pollution Control Federation Staff, eds., *Gravity Sanitary Sewers Design and Construction,* American Society of Civil Engineers, New York, 1982.

Brater, E. F., and Horace King, *Handbook of Hydraulics,* 6th ed., McGraw-Hill, New York, 1976.

Cole, Jonathan H., and Stephen F. Torchia, *Defying Gravity, CE News,* February 1998, pp. 67–69.

Dewberry & Davis, *Land Development Handbook,* McGraw-Hill, New York, 1996.

Intellisolve, *Hydraflow* (brochure), Marietta, Georgia, 1992.

Intergraph Corporation, *Insewer* (brochure), Huntsville, Alabama, 1992.

Metcalf & Eddy, Inc., *Wastewater Engineering,* 3d ed., McGraw-Hill, New York, 1990.

National Clay Pipe Institute, *Clay Pipe Engineering Manual,* Washington, D.C., 1982.

9

Storm Drainage

On every project, attention must be given to the impact of storm water. Precipitation falls on the site in the form of rain, hail, or snow, and precipitation runoff from other areas flows through the project. Storm drainage facilities must be designed to protect people and property from storm water inundation. Designing storm drainage systems requires an understanding of *hydrology* (the science of the natural occurrence, distribution, and circulation of the water on the earth and in the atmosphere), *hydraulics* (the science of the mechanics of fluids at rest and in motion), and drainage law. Understanding the elements of the design of storm facilities and their coordination with surface improvements and underground utilities is essential. Drainage law varies from location to location and from time to time, so local drainage laws must be investigated and applied.

Hydrology

A formal study of hydrology includes complicated concepts of weather forecasting, storm water runoff, and stream flow routing, as well as the determination of groundwater characteristics. Fortunately, however, for the small areas that are ordinarily involved in land development, the rational formula provides a conservative flow rate that can be used for designing storm water facilities. For projects where the drainage basin affecting the project is larger than 120 ha (320 acres), hydrologists should be brought in and other methods used. Design of flood control projects will not be discussed in this book.

The rational formula is given in Eq. 9.1.

$$Q = kCIA \qquad (9.1)$$

where Q = flow rate (m³/s or cfs)

$$k = \text{a factor to account for units}$$

S.I. $0.00278 \text{ m}^3\text{/s, per ha, mm/h}$

Imperial $1.008 \text{ cfs, per ac, in/h}$

$C = \text{runoff coefficient}$

$I = \text{rainfall intensity (mm/h or in/h)}$

$A = \text{area (ha or ac)}$

Rainfall intensity

The intensity factor I in the rational formula is the rate of rainfall over the area in mm per hour. This factor can be taken from an intensity-duration-frequency (IDF) chart (Fig. 9.1) supplied by the responsible agency, or the weather bureau may provide an IDF chart. To find the intensity from the chart, you must know the return period. If the return period is 100 years, the rate of rainfall given is of the most intense storm expected during a 100-year period. This is called a *100-year storm* or the *100-year event*. If the return period is 10 years, the rate of rainfall is of the most intense storm expected to occur during a 10-year period and is called a *10-year storm*. The larger the interval, the greater the intensity. The jurisdiction responsible for flood control will dictate what return period to use.

The duration (D) is the amount of time it takes for a drop of rain to travel from the most distant point in the drainage basin (described later) to the drainage structure. This is called the time of concentration t_c. There are complicated formulas to determine the time of concentra-

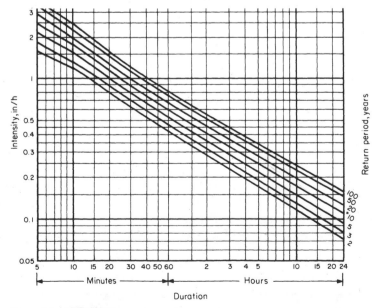

Figure 9.1 IDF Chart.

tion. One that is used to approximate the time of concentration for a pear-shaped basin is the Kirpich equation, which can be used for the time of concentration for overland plus channel flow.

$$t_c = 0.0078 \left(\frac{L}{S^{0.5}} \right)^{0.77} \tag{9.2}$$

where t_c = time of concentration (min)

L = horizontal projected length of the watershed (meters or feet)

$S = H/L$, where H is the difference in elevation between the most remote location in the watershed and the point of concentration (meters or feet) and L is the horizontal length between those same two points.

For the small-scale hydrologic problems involved in land development projects, an estimate of the duration is adequate. Ten minutes is commonly used as a minimum time for the overland flow to reach a swale where the flow is concentrated. If the distance is from the rooftop to the curb and gutter, 10 to 15 min can be used. Here, the most distant point in the drainage basin is judged to be the rooftop. The gutter is the drainage structure. The time of concentration increases as the drop of rain continues downstream.

When you have the return period and duration, the intensity can be read from an IDF chart. To determine the intensity from Fig. 9.1, if the return period is 5 years, find it on the right-hand side of the chart. The 5 is at the end of a diagonal line. Now, find the duration of 15 min at the bottom of the chart. Follow the vertical line representing 15 min until it intersects the diagonal line for the 5-year return period. The intersection falls about halfway between the horizontal lines for 1 and 1.5 in/h. The resulting intensity is 1.25 in/h. Notice that as the duration becomes longer, the intensity diminishes. The reason for the decrease in intensity is that peak intensity is seldom sustained for long. The average intensity is less for longer periods of time.

Storm water runoff

The runoff coefficient C in the rational formula (Eq. 9.1) represents the amount of water running off as a proportion of the total amount of precipitation falling on the area. Of the precipitation that reaches the ground, some will percolate into the soil, some will be taken up by the vegetative cover, some will evaporate, and the remainder will run off. For streets, runoff coefficients range from 0.70 to 0.95. That is, 70 to 95 percent of the precipitation falling on the area will run off. The responsible agency may provide a table of coefficients to use. Table 9.1 can be used when the design frequency is 5 to 10 years. The coefficient reflects the type of soil, type of ground cover, and the evenness and degree of slope. Typically, the area will consist of more than one type of cover, and a weighted average should be used.

**TABLE 9.1   Typical Runoff Coefficient (*C*) Values
for 5- to 10-Year Frequency Design**

Description of area	Runoff coefficients
Business	
Downtown areas	0.70–0.95
Neighborhood areas	0.50–0.70
Residential	
Single-family areas	0.30–0.50
Multiunits, detached	0.40–0.60
Multiunits, attached	0.60–0.75
Residential (suburban)	0.25–0.40
Apartment dwelling areas	0.50–0.70
Industrial	
Light areas	0.50–0.80
Heavy areas	0.60–0.90
Parks, cemeteries	0.10–0.25
Playgrounds	0.20–0.35
Railroad yard areas	0.20–0.40
Unimproved areas	0.10–0.30
Streets	
Asphaltic	0.70–0.95
Concrete	0.80–0.95
Brick	0.70–0.85
Drives and walks	0.75–0.85
Roofs	0.75–0.85
Lawns, sandy soil	
Flat, 2%	0.05–0.10
Average, 2 to 7%	0.10–0.15
Steep, 7%	0.15–0.20
Lawns, heavy soil	
Flat, 2%	0.13–0.17
Average, 2 to 7%	0.18–0.22
Steep, 7%	0.25–0.35

SOURCE: Warren Viessman, Jr., Terrence E. Harbaugh, and John W. Knapp, *Introduction to Hydrology,* Intext, New York, 1972, p. 306.

Example 9.1   Determine the runoff coefficient for an area that is 65 percent paving and buildings and 35 percent landscaping.

solution

1. Obtain the runoff coefficients from Table 9.1 for paving (streets), $C = 0.95$, and landscaping (lawns), $C = 0.22$.

2. Calculate a weighted average.

C		Portion of area		Total
0.95	×	0.65	=	0.62
0.22	×	0.35	=	0.08
		C	=	0.70

When downstream drainage facilities are inadequate, detention ponds can be used to mitigate the problem. Detention ponds are small-scale flood control reservoirs. The purpose of these ponds is to catch and detain storm runoff. Ordinarily, the pond is designed so that the amount of runoff that the downstream drainage facilities can accommodate is released, while the excess runoff is held back and released at a controlled rate. Software for the calculation of the capacity necessary for a detention pond is available, but the calculations should be performed only by engineers with experience in detention pond design or by a hydrologist. Design of a detention pond is beyond the scope of this book.

Determination of the area

The area (A) in the rational formula (Eq. 9.1) is the area of the drainage basin. A drainage basin or watershed is that area of land from which drainage contributes to a particular waterway. When the waterway referred to is a river, the area is called a *river basin*. A river basin is made up of small, tributary drainage basins. Several drainage basins are illustrated in Fig. 9.2. Ridges W, X, Y, and Z and swales A, B, and C are shown. Drainage basin A is bounded by ridges W, X, and Y and contributes storm water to swale A. Drainage basin B is bounded by ridges Y and Z; water falling there contributes to swale B.

To determine the amount of runoff reaching the point of concentration at A, delineate the drainage basin contributing to that point in the swale (waterway). Water flowing overland follows the steepest route. The flow line of the steepest route will always be perpendicular to the contours. The land is steepest where the contours are closest. To delineate the drainage basin contributing to a particular point, trace the flow line from point A up the contours at right angles (Fig. 9.2).

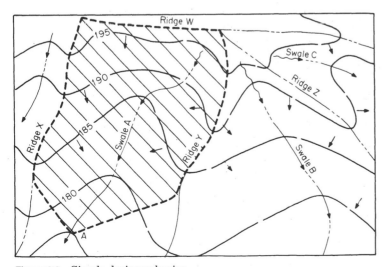

Figure 9.2 Simple drainage basins.

For most projects, the specific project topography will not be sufficient. USGS maps are typically used to determine the drainage basins which will impact the project (Fig. 9.3). Be aware that the rational formula is considered sufficiently precise only for small drainage basins of 120 ha (320 acres) or less. If the drainage basin is larger than 120 ha, a hydrologist should determine the quantity of runoff to expect.

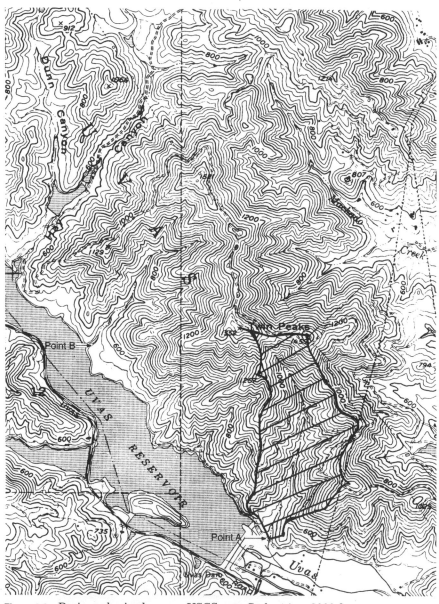

Figure 9.3 Drainage basin shown on USGS map. Scale: 1 in = 2000 ft.

Drainage basins in a developed area are shown in Fig. 9.4. One drainage area is bounded on the north by the crown on A Street, on the south by the lot line between lots 2 through 5 and 8 through 11, and on the east and west by ridges through lots 1 and 6. This drainage area is collected at catch basin A (CB A). Catch basin B collects water from the area bounded by the ridges previously described through lots 1 to 11 and by the crowns on First Street, Second Street, and B Street.

The first step in drainage system design is to develop the grading plan (Chap. 6). On-site surface drainage basins are created to direct runoff to ditches and storm water inlets. Six of the lots shown in Fig. 9.4 will interface with existing drainage basins along the northerly tract boundary. In this case, the lots will be graded so that the northern half of each lot drains north and the southern half of each lot drains to A Street. First and Second Streets slope south. Two of the drainage basins established when the lots are constructed this way are delineated in Fig. 9.4. The storm water falling on the basins will collect in the ditch along the northerly tract boundary and be picked up by field inlets (FI) 1 and 2.

The ditch here must be designed to accommodate the off-site drainage basins to the north as well. Notice that the runoff flows perpendicular to the contour lines, as shown by the flow lines. At the northwest corner of the tract, the basin is limited by the point from which the

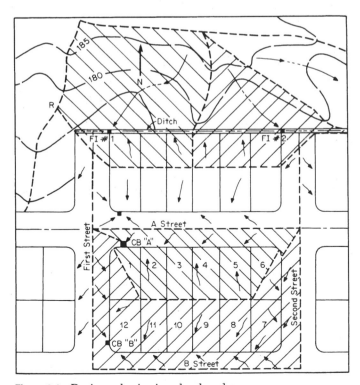

Figure 9.4 Drainage basins in a developed area.

water flows. Water falling south of point R flows south and will not reach the site. Therefore, the limit of the basin is as shown. Once the boundaries of the drainage basins have been defined, their areas can be calculated. If the drainage basin is irregular, use a planimeter as described in Chap. 6. Convert the area to hectares (acres) before putting it into the rational formula. Once the quantity of runoff Q has been established, the size and type of drainage facilities can be designed.

Figure 9.5 shows the drainage basins established on a highway within the right-of-way. These drainage areas must be delineated and measured to determine the runoff from the highway. Drainage will run from the high point in the profile to the low point and from the high side of the cross section to the low side, where a drop inlet or other drainage facility must be located.

Cross Drainage

For linear projects in rural areas, hills are often cut down and valleys filled in to create usable grades for the projects. When this is done, the fill material cuts off natural waterways. Runoff in these channels must be provided access to continue its course, so culverts and bridges are included in the design. For simplicity, linear projects will be referred to as highways in this section, but the discussion is meant to refer to any linear project that involves placing embankment material across waterways.

A number of considerations affect the solution to cross-drainage problems. Some questions that must be answered are as follows:

Will a bridge be needed, or will a culvert be adequate?

If a box culvert is used, should it be a rectangular or a square box culvert?

If a pipe is used, should it be concrete or corrugated metal?

Should the pipe be circular, elliptical, or arched?

Will head walls and end walls be needed?

Should entrance structures be used to improve capacity or protect embankments?

At what return period is it acceptable for the runoff to overtop the highway?

How much freeboard should be used?

Will debris racks be needed?

Will energy dissipaters be needed?

Will slope protection be needed?

What material should be used for the culvert to prevent corrosion by the soil?

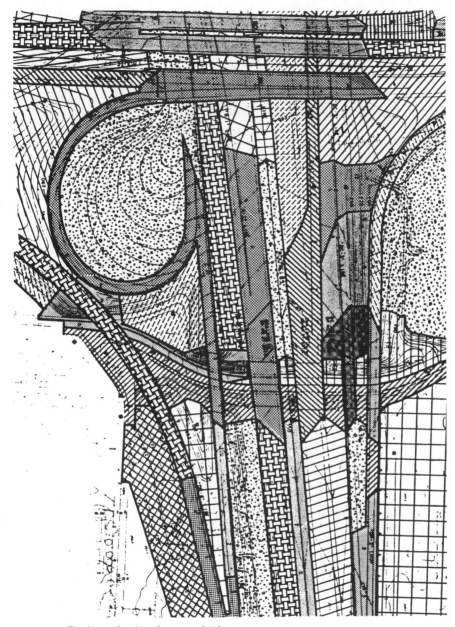

Figure 9.5   Drainage basins shown on highway.

Each of these questions must be answered while keeping in mind its impact on safety, costs, maintenance, and, in some cases, aesthetics.

The hydraulics involved in sanitary sewer systems and underground storm drain networks are of uniform flow, and capacities can be determined with relative ease. However, because of the short lengths of highway crossings, flows there are usually nonuniform, and the hydraulics of culvert and bridge design are, therefore, extremely complex. The capacities of the culverts are much lower than could be expected from uniform flow. A number of culvert-design software programs are available commercially. The Federal Highway Administration (FHWA) describes culvert hydraulics in *Hydraulic Design of Highway Culverts,* Hydraulic Design Series No. 5, and there are a number of textbooks that cover the subject. However, calculation of culvert design and the use of that software should be limited to those engineers who understand the complexities of culvert hydraulics. Cross-drainage design should be performed only by engineers with extensive experience in hydraulics.

Surface Improvements and Structures

The most important surface improvement for controlling storm water is grading. The earth must be shaped to direct water safely to ditches and storm water inlets, from which it can be carried away. The design of grading to direct drainage is described in Chaps. 6 and 7.

Ditches

When runoff from an unimproved area reaches a site as overland flow, it must be intercepted and collected. This can be accomplished with ditches. As with sewers, the amount of runoff determines the design of the ditch.

Unlined ditches should be considered. When the runoff being handled is very small, and the ditch is less than 30 m (100 ft) long, a simple note, "Grade To Drain," at the flow line of the ditch on the plan may be sufficient for construction. Where the volume of runoff is low, slopes should be at least 1 percent. A shallower slope may become uneven in time.

An unlined ditch with a slope that is too steep will erode and can threaten the roadway or other improvements. The maximum allowable slope depends on the volume of runoff and the type of soil. If the soil is sandy, the maximum limit for the slope of an unlined ditch should be 0.025. If the soil is compacted clay and the flow is less than 0.03 m³/s (1 cfs), the slope can be as high as 0.06.

Higher volumes of runoff will require lining the ditch. Where erosion will be a problem, the ditch can be lined with any of a number of materials, such as asphalt, concrete, Gunite, or cobblestone. Economics and velocities will indicate which choice is best. A minimum slope of 0.003 should be used for a paved gutter for roadway runoff in an urban setting. Successful construction of a flatter slope is doubtful.

The cross section of the ditch must be designed to fit the circumstances and accommodate the flow (Fig. 9.6). A V ditch is most economical to build. If the ditch is located where people are likely to step into it, a shallow, flat-bottomed, or curved ditch is better. If the ditch is to carry a large volume of runoff, a trapezoidal ditch is more efficient. Designing the shape of the ditch will be an iterative process. You select a cross-sectional area and calculate the capacity using the continuity equation and Manning's equation. After comparing the designed capacity to the required capacity, you redesign to provide greater capacity or a more economical design (see Fig 9.7). To determine the required cross-sectional area of the ditch, use the continuity equation in the form of Eq. 8.4, repeated here as Eq. 9.3.

$$A_R = \frac{Q}{V} \qquad (9.3)$$

where A_R = cross-sectional area required (m^2 or ft^2)
    Q = quantity (m^3/s or cfs)
    V = velocity (m/s or fps)

Q is determined from Eq. 9.1. Velocity V is determined from Manning's equation, Eq. 8.5, repeated here as Eq. 9.4.

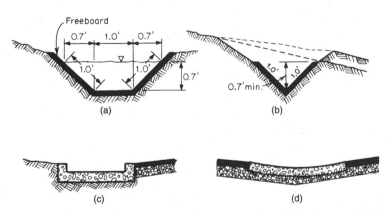

Figure 9.6 Types of ditches: (a) trapezoidal; (b) V ditch; (c) flat-bottomed; (d) curved-bottomed.

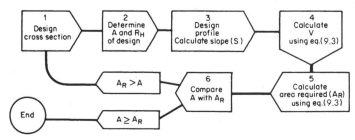

Figure 9.7 Flow chart for ditch design.

$$V = \frac{R_H^{2/3} S^{1/2}}{n} \tag{9.4}$$

where n = coefficient of friction

$$R_H = \text{hydraulic radius, } \frac{\text{area}}{\text{wetted perimeter}} \text{ or } \frac{a}{p}$$

S = slope (m/m or ft/ft)

The responsible jurisdiction may have a table of values of n, a surface roughness factor. Otherwise, use Table 9.2. The hydraulic radius (R_H) is the cross-sectional area of the ditch (a) divided by the wetted perimeter (p). The wetted perimeter is the length of the surface on the cross section that will be wet when the ditch is at design capacity (see Example 9.2). The slope is that of the ditch profile.

Example 9.2   Calculate the hydraulic radius R_H for Fig. 9.6a.

solution

1. Add the lengths of the surfaces on the cross section that will be wet when the ditch is flowing at design capacity.

$$p = 1 \text{ m} + 1 \text{ m} + 1 \text{ m} = 3 \text{ m}$$

2. Calculate the cross-sectional area of the ditch when it is flowing at design capacity.

$$a = \frac{b_1 + b_2}{2} n$$

$$= \frac{(0.7 + 1.0 + 0.7) + 1}{2} \times 0.7 \text{ m}$$

$$= 1.19 \text{ m}^2$$

3. Calculate R_H.

$$R_H = \frac{a}{p} = \frac{1.19}{3} = 0.40$$

The design of the ditch may be shown entirely on the cross section by showing a minimum depth below existing ground for the flow line of the ditch, as shown in Fig. 9.6b. The grading contractor can then cut the ditch without needing survey stakes for vertical control. If the design requires more exact vertical control, the profile (flow line) elevations should be shown on the grading plan or the plan view of the construction plans at grade breaks. This way, you may be able to avoid

TABLE 9.2   Values of *n* to Be Used with Manning's Equation

Surface	Best	Good	Fair	Bad
Uncoated cast-iron pipe	0.012	0.013	0.014	0.015
Coated cast-iron pipe	0.011	0.012*	0.013*	
Commercial wrought-iron pipe, black	0.012	0.013	0.014	0.015
Commercial wrought-iron pipe, galvanized	0.013	0.014	0.015	0.017
Polyvinyl chloride (PVC) pipe	0.009	0.010	0.011	
Smooth brass and glass pipe	0.009	0.010	0.011	0.013
Smooth, lockbar and welded "OD" pipe	0.010	0.011*	0.013*	
Riveted and spiral steel pipe	0.013	0.015*	0.017*	
Vitrified sewer pipe	{0.010⎫			
	0.011⎭	0.013*	0.015	0.017
Common clay drainage tile	0.011	0.012*	0.014*	0.017
Glazed brickwork	0.011	0.012	0.013*	0.015
Brick in cement mortar; brick sewers	0.012	0.013	0.015*	0.017
Canals and ditches				
Earth, straight and uniform	0.017	0.020	0.0225*	0.025
Rock cuts, smooth and uniform	0.025	0.030	0.033*	0.035
Rock cuts, jagged and irregular	0.035	0.040	0.045	
Winding sluggish canals	0.0225	0.025*	0.0275	0.030
Dredged earth channels	0.025	0.0275	0.030	0.033
Canals with rough stony beds, weeds on earth banks	0.025	0.030	0.035*	0.040
Earth bottom, rubber sides	0.028	0.030*	0.033*	0.035
Natural stream channels				
1. Clean, straight bank, full stage no rifts or deep pools	0.025	0.0275	0.030	0.033
2. Same as 1, but some weeds and stones	0.030	0.033	0.035	0.040
3. Winding, some pools and shoals, clean	0.033	0.035	0.040	0.045
4. Same as 3, lower stages, more ineffective slopes and sections	0.040	0.045	0.050	0.055
Neat cement surfaces	0.010	0.011	0.012	0.013
Cement mortar surfaces	0.011	0.012	0.013*	0.015
Concrete pipe	0.012	0.013	0.015*	0.016
Corrugated metal pipe	0.025*	0.025*	0.025*	0.025*
Wood stave pipe	0.010	0.011*	0.012	0.013
Plank flumes				
Planed	0.010	0.012*	0.013	0.014
Unplaned	0.011	0.013*	0.014	0.015
With battens	0.012	0.015*	0.016	
Concrete-lined channels	0.012	0.014*	0.016*	0.018
Cement-rubble surface	0.017	0.020	0.025	0.030
Dry-rubble surface	0.025	0.030	0.033	0.035
Dressed-ashlar surface	0.013	0.014	0.015	0.017
Semicircular metal flumes, smooth	0.011	0.012	0.013	0.015
Semicircular metal flumes, corrugated	0.0225	0.025	0.0275	0.030
5. Same as 3, some weeds and stones	0.035	0.040	0.045	0.050
6. Same as 4, stony sections	0.045	0.050	0.055	0.060
7. Sluggish river reaches, rather weedy or with very deep pools	0.050	0.060	0.070	0.080
8. Very weedy reaches	0.075	0.100	0.125	0.150

* Values commonly used in designing.

SOURCE: Adapted from E. F. Brater and Horace King, *Handbook of Hydraulics,* McGraw-Hill, New York, 1976, pp. 7–22.

drawing the profile on the finished plans. However, the engineer should draw the existing ground and profile and perform the necessary calculation to verify that the ditch will perform as needed.

To design the ditch profile, draw a ground line profile at the centerline or edges of the ditch. Draw a line roughly parallel with and below the lowest ground line profile (Fig. 9.8). The ditch profile must be below the ground at least as much as the ditch is deep. That is, if the ditch is 0.3 m (1 ft) deep, the flow line profile must be at least 0.3 m (1 ft) below the natural ground everywhere at the edge. Otherwise, the ditch will come out of the ground. Make no more breaks in the profile than are necessary to accommodate the changes in the ground line profile. If the cross slope is steep or erratic, it may be necessary to draw cross sections at critical points to verify that catch points will be within the property or within a reasonable distance. When the ditch profile is drawn, the slopes must be calculated all along its length. For each section of the profile, divide the difference in elevations at the beginning and end of the section by that length of the section. Continue these calculations until you have established the slopes all along the profile. Keep in mind that, if the slope changes from a steep slope to a flatter slope, you my have to add some freeboard to account for possible hydraulic jump.

When a slope has been selected, use it in Manning's equation (Eq. 9.4) to calculate V. Then divide V into Q (Eq. 9.3) to get the cross-sectional area required. Compare the area required with the area the ditch section provides. If the cross-sectional area of the ditch is larger, the design will work; if not, try using a larger cross section and/or a steeper slope. Then go through the procedure again using the new R_H and/or S (Fig. 9.7). If the ditch is long or lined and the section is much larger than necessary, it is advisable to make the cross section smaller, and thus cheaper. Here again, if the design of the cross section is changed, a new cross-sectional area and a new wetted perimeter must be determined and R_H recalculated.

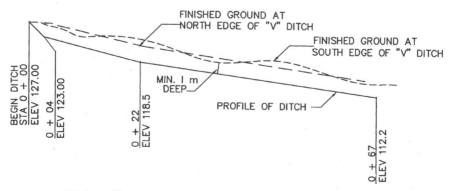

Figure 9.8   Ditch profile.

Storm water inlets

At the low point in the ditch or along the gutter in roadways, the runoff is collected and routed underground or discharged into an approved waterway. To collect the runoff for removal in an underground system, a storm water inlet (SWI) is used (Fig. 9.9). Inlets are also referred to as drop inlets (DI), flat grate inlets (FGI) or catch basins (CB). When placed at the street edge, a hooded catch basin which becomes part of the curb is usually used. If the inlet is going to be on public property, the approving public agency should have standard plans showing what to use. On private property, choose an inlet from a catalog of locally available, prefabricated products whenever possible. Identifying the inlet as "Christy Inlet No. 241 or equivalent" gives the contractor some flexibility. If the size and shape of the ditch will not accommodate the inlet chosen, show a transition to a cross section that fits the inlet.

Verify that there is enough space in the walls of the inlet to accommodate the pipes entering and leaving. Where the pipe enters on the skew, make a large-scale 1:20 (1 in = 2 ft) sketch of the plan view. If two pipes enter the same side of an inlet, a cross-sectional sketch may be needed.

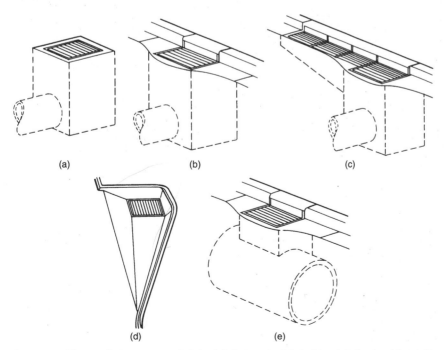

Figure 9.9 Types of storm water inlets: (*a*) flat-grate inlet; (*b*) catch basin; (*c*) catch basin with gallery; (*d*) offset inlet; (*e*) catch basin over large storm main.

Manholes

A cross section of a typical manhole is shown in Fig. 8.5. Note the eccentric cone on top. This serves to reduce the opening diameter so that a 0.6 m (2-ft) diameter manhole cover can be used over a 1.2 to 1.5 m (4- to 5-ft) diameter manhole. The vertical side of the cone should be oriented so that when workers use the ladder, they will be facing oncoming traffic. Where the manhole is located directly under some obstacle on the surface, such as on the curb and gutter, the cone can be rotated away from the obstacle. The tops of pipes on incoming lines must be below the cone. If the line is unusually shallow and this is not possible, a custom manhole must be designed. When an incoming line enters with the invert more than 0.76 m (2.5 ft) above the bottom of the manhole, an outside drop (Fig. 8.5) should be used.

Some manholes will have several lines entering. To verify that there is sufficient space in the walls of the manhole for pipes to enter without intersecting other pipes, prepare a plan-view drawing and a cross section at a scale of 1:20 (1 in = 2 ft). If the drawings show that there is not enough space, the manhole must be relocated or another manhole added.

The Storm Drain Network

Complicated criteria are applied to the design of storm drain piping systems: agency requirements, the physical laws of hydraulics, coordination with new and existing facilities, construction technology, and cost considerations. The agency may dictate criteria such as horizontal locations, depth, velocity of flow and minimum slopes, minimum pipe sizes, types, and classes. The agency should have a master storm drainage plan of all the storm runoff requirements in the area and may dictate pipes large enough to accommodate future growth. Usually, construction specifications and standard plans of facilities are dictated.

Locating inlets

Catch basins should be located at the low points in streets, at the low points of intersections (usually at one end of curb returns), and at intervals to satisfy the other criteria stated previously. If the criterion is a specified width of gutter flow, treat the gutter the same as a ditch. The available gutter width determines the dimensions of a modified V ditch. The dimensions and cross-sectional area are found on the required street section. The longitudinal slope is the street's slope. An interval may not be stated directly, but rather will be a matter of the reach of the flow into the traveled way. Drainage backing up along a roadway can be a serious hazard to traffic safety. Ordinarily, runoff along the edge of a roadway between the edge of the traveled way and the curb or dike (the shoulder area) is acceptable. But when the flow is deep enough to reach beyond the shoulder into the traveled way, it must be directed into an inlet. When the allowable overland flow reaches the limiting criterion, the runoff must be picked up. Ordinarily, pickup is made with a catch basin. All the locations of catch basins must be determined before the sewer network can be laid out. Where the catch basin is in a sag curve, one or two flanking inlets should be provided and a determination of the drainage release point made in case the inlet fails to function. The extent of resulting flooding should be determined.

Capacities of inlet openings must be determined. The capacity of the inlet depends on the cross slope, the longitudinal slope, and the depth of flow at the limiting condition, and on whether the location is in a continuous longitudinal slope or at the low point of a sag curve. Because there is significant risk of an inlet being blocked by debris, it may be wise to install more than one inlet at a sag-curve location. Highway Engineering Circular #12 (HEC-12), Drainage of Highway Pavements, describes the methods of determining capacities under various conditions. The circular is available through the National Technical Information Service (NTIS) in Springfield, Virginia.

On private commercial, industrial, and multifamily residential sites, as well as on public streets in some jurisdictions, storm water inlets rather than manholes are used as junction boxes. In this case, inlets or storm drain clean outs should be spaced no more than 30 m (100 ft) apart so that maintenance can be performed on the pipes unless the diameter of the pipe is larger than 0.76 m (2.5 ft). The conduits must be deep enough so that the tops of the laterals at the storm water inlet have enough cover. If the end of the lateral is in a paved area, the top of the lateral should be below the subgrade. If the lateral is in a landscaped area, it should be at least 0.5 m (1.4 ft) deep, so that gardening equipment will not damage it. Each downstream line must be larger than the previous line, regardless of hydraulic requirements, to prevent clogging of the line.

Some agencies allow the use of valley gutters (Fig. 9.10). The valley gutter transfers the gutter flow from one side of a secondary street to the other, thereby eliminating the need for at least one catch basin and some storm drain lines. When a valley gutter is used, set the elevations at the points marked on Fig. 9.10 to ensure that there will be no ponding.

Coordinating facilities

When the grading plan has been designed and ditches and catch basins have been located, the underground network of storm drains can be laid out. The storm lines must be coordinated with the sanitary sewer system as well as with the grading and other existing and proposed utilities, so you need to again use the project master plan. All facilities,

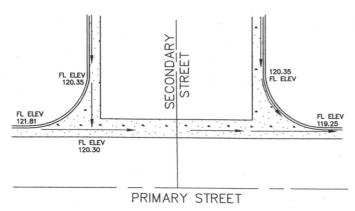

FL ELEV
120.35

SECONDARY STREET

120.35
FL ELEV

FL ELEV
121.81

FL ELEV
119.25

FL ELEV
120.30

PRIMARY STREET

Figure 9.10   Plan view of concrete valley gutter.

new or existing, that will require accommodation should be shown on the master plan as described in Chap. 8. The proposed sanitary sewer lines should have been drawn on earlier.

To design the storm drain system, all available outfalls must be identified. An *outfall* is a place where a storm drain can discharge. It may be a river, channel, stream, lake, or, most commonly, another storm drain line. To be available, the outfall must have unused capacity and an invert elevation low enough to accommodate the new system. Write the invert elevations on the master plan to use later to coordinate crossings. The capacity required will be based on the runoff from all the drainage basins that contribute to it.

When laying out and coordinating the storm drain and sewer systems, keep in mind what will be most economical and most expedient. Plot the locations and elevations of all the storm and sanitary outfalls that are available. When more than one outfall is used, the total length and required diameter of resulting mains may be less. Plotting all the outfalls will indicate what flexibility there is. Decide whether the storm drain or the sanitary sewer system has more critical restraints. If it is not clear which line is more critical, the sanitary system should be laid out first.

Some agencies dictate the horizontal locations of storm and sanitary lines, for example: "Storm mains must be installed 1.5 m (5 ft) north or east of the centerline of the street." If the horizontal location of the storm drain line is not specified, consider what will be most economical. First, it must not interfere with the sanitary line or other utilities. If there are more storm laterals on one side of the street than on the other, place the storm line on the side with more laterals. That way, more of the laterals will be shorter from the property line to the main.

Consider locations that minimize crossings. In Fig. 9.11, the sanitary mains are located on the north and east sides of the streets and the storm mains are located on the south and west sides. There is a cross-

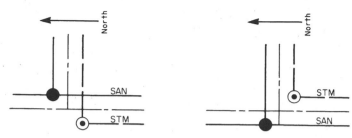

Figure 9.11   Horizontal layout of sewers.

ing point in the southeast quadrant. In this example, if the storm main is positioned south and east and the sanitary main is positioned north and west, that crossing can be eliminated (Fig. 9.11). Not only is there then one less crossing clearance to calculate, but, more important, there is one less potential trouble spot if there is a design or construction discrepancy.

When laying out the storm main on the plan view, curves are sometimes allowed. The allowable degree of deflection or radius of curvature is specified by the pipe manufacturer. Some agencies do not allow curved mains; others do under specific circumstances, such as where a manhole is provided at the BC and/or EC. When curves are to be used, exact locations of the BC and EC (see Chap. 7) as well as curve data must be calculated before the design is complete.

Designing the Storm Drain Profile

When the horizontal layout is complete on the master plan, design of the profile can begin. The actual design will be done on the profile worksheet described in Chap. 8 where the proposed sanitary sewer line is already shown. The profile should show all existing and proposed crossings: underground electrical, gas, communication, television cable, and water lines. It is not unusual for there to be more than one conduit for each utility. At crossings, both the tops and the bottoms of conduits should be shown. In streets that have been used for many years, the utilities could form a vertical curtain along the street. Because the storm drain line is a gravity flow line, it is particularly important that a thorough search be made to find all underground utilities or other obstacles to the main and its appurtenances. Show critical points, such as utility crossings, and design the profile to clear them. The existing and proposed ground line should be plotted as well.

If you are working in an area that is already developed, there may be little flexibility as to where the profile will go without impacting existing lines. You may have to thread the storm drain through existing lines at street crossings. Those areas with the least flexibility must be designed first. If you cannot avoid conflict with an existing line, those lines which are not dependent on gravity flow can be moved, but that alternative should be avoided if possible because it is expensive.

Where existing utilities are not a problem, begin from the outfall and project the profile upstream. Use inverts on the storm drain that are sufficiently above the sanitary lines to allow the bottom of the storm line to clear the top of the sanitary lateral. Some agencies require a minimum amount of clearance. The inverts of the mains must be deep enough so that the tops of the storm drain laterals will be below the street subgrade at the back of the curb or edge of the pavement (Fig. 9.12). This serves two purposes: by putting the pipes below the roadwork, they are protected from being damaged by construction equipment; and the subgrade distributes the traffic loads, protecting the pipes from being crushed.

If the sanitary main is shallow or the storm line is larger than 0.76 m (30 in), it may be necessary to design the storm line profile below the sanitary line (Fig. 9.13). Here, storm laterals must clear the sanitary main. If manholes are required at the storm lateral connections, the storm laterals can then go over the sanitary main and clearance should not be a problem.

The class of pipe to use should be selected from the manufacturer's specifications based on the character of the expected traffic loads and trench depths. A concrete cap can be poured over pipes where additional protection is needed for shallow pipes or where pipe crossings have little clearance (Fig. 9.14). Cast iron pipe (CIP) or ductile iron (DI) pipe can be used in extreme cases.

When the profile changes horizontal or vertical direction and when the pipe size changes, a manhole or storm water inlet is usually required. Storm water inlets are usually cheaper but, if they are deeper than 1 m (30 in), they must be made large enough for a maintenance worker to enter them safely to clear debris. A manhole or inlet is also required at intervals of 150 m (500 ft) or less to facilitate maintenance. However, when connecting to deep manholes or inlets, the invert should be no higher than 1 m (or 2.5 ft) above the bottom of the manhole, to avoid the cost of an outside drop.

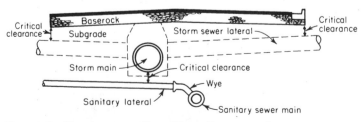

Figure 9.12   Street cross section with sanitary sewer deeper.

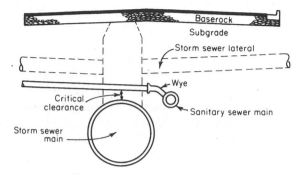

Figure 9.13 Street cross section with sewer deeper.

The agency may dictate vertical criteria, for example, "The top of pipe must be at least 0.6 m (2 ft) below the top of curb," or "The hydraulic grade line (HGL) must be a minimum of 0.15 m (6 in) below the top of curb for the 100-year storm." Hydraulic grade line is explained later. Identify critical points on the profile, such as crossings and changes of slope in the surface profile. Keep in mind the criteria stated earlier about laterals. Where no other factors govern, keep the pipe as shallow as possible in order to keep down the cost of trenching and the depths of manholes and inlets.

Ordinarily, the agency will dictate the range of acceptable flow velocities. The velocities determine minimum and maximum slopes. If the agency has not dictated velocities, use a minimum of 0.6 m/s (2 fps) and a maximum of 3 m/s (10 fps). In a 300 mm (12-in) diameter concrete pipe, a slope of 0.003 yields 0.6 m/s (2 fps); a slope of 0.064 yields 3 m/s (10 fps).

Determining pipe sizes

Calculation of the size of pipe required to accommodate a single drainage basin is simple. All that is needed is the amount of runoff reaching the pipe in cubic meters (or cubic feet) per second, a value to use for Manning's $n,$ and some hydraulics calculations. The method for

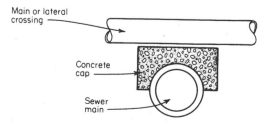

Figure 9.14 Concrete cap.

determining the runoff was explained earlier in this chapter; the value to use for Manning's n will be dictated by the responsible agency or can be taken from Table 9.2.

The choice of size and slope is made based on hydraulic factors, the criteria of the responsible agency, criteria affecting the design of the profile as described earlier, and the cost of material and trenching. Though providing for a single drainage basin is simple, providing for multiple basins served by a piping network is complicated. A storm drainage system calculations form is included here as Fig. 9.15. When filled out correctly, the pipe sizes and slopes and flow velocities are shown.

The calculations can be made with the use of a spread sheet as illustrated in Fig. 9.17. Once the form has been prepared it can be used repeatedly by simply replacing the data but not the formulas. Additional columns can be added for pipe inverts and cover as shown in Fig. 8.9. Hydraulic grade line calculations can also be included in additional columns. The only problem with making your calculations this way is that it is easy to accidentally corrupt one of the formulas without realizing it, so unusual care must be taken. Extra care must be taken as well to analyze the results visually to catch any errors.

Results of Example 9.3 are shown in Fig. 9.17 as calculated on a spread sheet. The beauty of this approach is that the engineer can easily experiment with pipe sizes and slopes to get different capacities. This allows the engineer to quickly pick the most efficient combination.

Filling out the form must be coordinated with design of the profile and may have to be done more than once to coordinate with other criteria. Filling out the form is complicated, so Example 9.3 is used for illustration. Note that, in this example, the time of concentration t_c is given as the time for a drop of rain to reach the storm water inlets, whereas t_c is often defined as the time for a drop of rain to reach the gutter. Normally the t_c will have additional flow time between where the drop of rain enters the gutter and where it enters the inlet. Flow time can be calculated using the equation

$$t = \frac{l}{60V} \tag{9.5}$$

where t = flow time (min)
  l = length (m or ft)
  V = velocity (m/s or fps)

The velocity of the flow in a concrete gutter can be taken from Table 9.3.

Example 9.3 Determine sizes, slopes, and inverts of the storm drain network shown in Fig. 9.16. The sizes of drainage basins, times of concentration, and coefficients of runoff are specified there. The agency criteria are as follows:

Minimum slope 0.004

Minimum pipe size 300 mm (12 in); 0.15 m (0.5-ft) drop at connections with a change of direction of 30° or more

Storm Drain System Calculations Form

Point of concen-tration	t_c (min)	C	I (mm/h)	A (ha)	Total Q (m³/s)	Diameter D (meters)	Circum-ference Wetted Perimeter	Pipe area (m²)	Hydraulic Radius (R_H)	Slope (m/100m)	Velocity (m/s)	Length (m)	Q=VA Capacity (m³/s)
A	B	C	D	E	F	G	H	I	J	K	L	M	N

Figure 9.15 Storm drainage calculations form.

TABLE 9.3   Approximate Mean Velocity of Flow in Gutters

Gutter slope	Velocity, m/s	Velocity, fps
0.004	0.6	1.9
0.006	0.7	2.4
0.008	0.85	2.8
0.010	0.9	3.1
0.015	1.15	3.8
0.020	1.3	4.4

Storm drainage systems calculations form (Fig. 9.15)

IDF chart (Fig. 9.1). Use inches/hour as mm/hr for this example.

10-year return period storm

Manning's $n = 0.015$ for pipes 300 to 840 mm (12 to 33 in); $n = 0.013$ for pipes 910 mm (36 in) or larger

solution

1. Fill in column A of the storm drainage systems calculations form. Start with the point of concentration farthest from the outfall (CB no. 1). Skip a line and list the next point of concentration (MH no. 1) on the third line. If there is more than one pipe emptying into the second point of concentration, skip a line on the form and list the other point of concentration upstream of the second point (CB no. 2). Continue through the network until all points of concentration have been listed (Fig. 9.17).

2. Fill in columns B, C, and E for the catch basins only. There is no overland runoff entering the manholes directly. All runoff reaching the manholes does so through pipes.

3. Using the t_c given in Fig. 9.16 and a 10-year return period, find the rainfall intensity at each catch basin from Fig. 9.1. Enter rainfall intensity in column 4.

4. The quantity of runoff (Q, column F) is determined using the rational formula (Eq. 9.1). Calculate Q for each catch basin by multiplying the values in columns C, D, and E. Enter the product into column F.

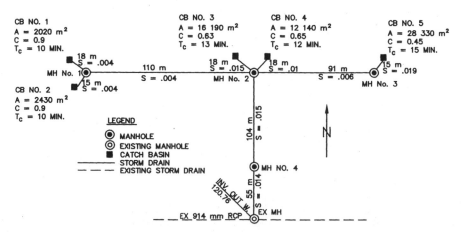

Figure 9.16   Storm drainage network for Example 9.3.

Point of Concentration	t	C	I mm/h	A (ha)	Total quantity required (m/s)	Diameter D (meters)	Circumference Wetted Perimeter R	Pipe Area (m)	Hydraulic Radius (R)	Slope (m/100m)	Velocity (m/s)	Length (m)	Q=VA Capacity (m/s)
A	B	C	D	E	F	G	H	I	J	K	L	M	N
CB#1	10	0.9	44	0.202	0.2288	0.305	0.96	0.073	0.076	0.4	7.52	18	0.55
MH#1	10.0												
CB#2	10	0.9	43	0.243	0.2690	0.305	0.96	0.073	0.076	0.4	7.52	15	0.55
MH#1	10.0												
CB#3	13	0.63	38	1.619	1.1085	0.381	1.20	0.114	0.095	1.5	16.90	15	1.93
MH#2	13.0												
CB#4	12	0.65	41	1.214	0.9253	0.381	1.20	0.114	0.095	1	13.80	18	1.57
MH#2	12.0												
CB#5	15	0.45	34	2.833	1.2397	0.381	1.20	0.114	0.095	1.9	19.02	15	2.17
MH#3	15.0												
MH #1	10	0.9	43	0.445	0.4925	0.305	0.96	0.073	0.076	0.4	7.52	110	0.55
MH#2	10.2												
MH #3	15	0.45	34	2.833	1.2397	0.460	1.44	0.166	0.115	0.6	12.12	91	2.01
MH#2	15.1												
MH #2	15.1	0.57	34	6.111	3.3161	0.533	1.67	0.223	0.133	1.5	21.16	104	4.72
MH #4	15.2	0.57	31	6.111	3.0235	0.533	1.67	0.223	0.133	1.4	20.44	55	4.56
EX MH#5	15.2	0.57	30	6.111	2.9259								

Figure 9.17 Storm drainage calculations form for Example 9.3.

5. Using Q for CB no. 1 and CB no. 2 and the value of n given (0.015), select a pipe size and slope and perform the calculations to verify that they work (see Example 8.2). The runoff reaching CB no. 1 is 0.23 m³/s (0.79 cfs). When the agency's minimum size pipe and slope are used, a flow of 0.55 m³/s (1.95 cfs) is accommodated. From this information, it can be seen that all pipes carrying 0.55 m³/s (1.95 cfs) or less can have a 300 mm (12-in) diameter. Fill in columns G through L for the catch basins on the line between the points of concentration. Notice that the slope is given in meters per 100 meters (feet per 100 ft). The slope (column K) will be taken from the profile. Slopes are given arbitrarily for this example.

6. The lengths of the pipes are determined from the plans. In this example, the lengths can be read from Fig. 9.16. Put the lengths of all the pipes into column M.

7. Column B is used for the amount of time it takes for the runoff to flow through the pipe between the points of concentration in column A. That time is calculated using Eq. 9.5.

$$t = \frac{l}{60 \text{ s/min } V}$$

For the pipe between CB no. 1 and MH no. 1,

$$t = \frac{18 \text{ m}}{60 \text{ s/min} \times 7.5 \text{ m/s}} = 0.04 \text{ min}$$

Add the additional time used for the runoff to move from the catch basin to MH no. 1 to the time of concentration. In this case, the 0.04 minutes is insignificant, but on longer runs the time of concentration will increase and a new rainfall intensity will have to be used for each new time of concentration. Go through the same procedure for all the catch basins.

8. Enter the values of C and A (the total tributary area) for each manhole. At MH no. 3, there is only one pipe entering, and the values of C and A are the same as at the contributing catch basin. When more than one pipe enters a manhole, a weighted average is used for the value of C.

From	A		C		
CB no. 1	2020 m²	×	0.9	=	1818
CB no. 2	2430 m²	×	0.9	=	2190
	4550				4008

At MH no. 1, $C = \dfrac{4008}{4550} = 0.9$

From	A		C		
CB no. 3	16 190 m²	×	0.63	=	10 200
CB no. 4	12 140 m²	×	0.65	=	7 890
MH no. 1	4 450 m²	×	0.90	=	4 005
MH no. 3	28 330 m²	×	0.45	=	12 750
	61 110				34 850

At MH no. 2, MH no. 4, and Ex MH,

$$C = \frac{34\ 850}{61\ 110} = 0.57$$

The value of C thus derived should be entered on the form for each manhole. The value of C is entered only once for each manhole.

9. Next, determine the longest t_c required for the runoff to reach the manhole farthest from the outfall. The t_c at CB no. 1 is 10 min. The flow time from CB no. 1 to MH no. 1 is 0.03 min. The t_c at CB no. 2 is 10 min. The flow time from CB no. 2 to MH no. 1 is 0.02 min. The total t_c through CB no. 1 is 10.03 min. The total t_c through CB no. 2 is 10.02 min. Fractions of a minute are not significant so are dropped.

10. Proceed by filling in the information for each MH. The amount of flow being picked up at CB no. 5 exceeds the capacity of minimum conditions. There is a range of sizes and corresponding slopes from which to choose. A pipe 380 mm (15 in) in diameter is the next larger size that will handle the 1.1 m³/s, so that size pipe is used. The choice should be based on the factors discussed earlier, in the section on profile design.

11. The t_c at MH no. 2 must correspond to the single longest amount of time necessary for runoff to reach that manhole. In this case, it takes 15.1 minutes for the runoff to travel from the farthest point in the basin to MH no. 2. Therefore, 15 minutes is the time of concentration used to determine the rainfall intensity for calculations of runoff from that point. (Making sense of the reasons for using this complex method requires an understanding of hydrologic concepts that is beyond the scope of this book.) Complete the form for flow from MH no. 2 to the existing manhole.

12. Now the invert elevations can be set. The exact invert elevation of the outfall should be available from survey notes. When sewers are connected, the inside tops of the pipes are matched—not the inverts. Where there is a change of pipe sizes, there is also a change of invert elevations. A drop is also required where there is a change of direction. Starting at the outfall, work up the network of drainage pipes.

Further information can be added to the form as needed. The differences between the inverts and the tops of the pipes for the various pipe sizes are available from the manufacturers. When the calculation of the profile is complete, critical points should be checked.

Example 9.4   Calculate the amount of cover over the surface low point at STA 8+78 in Fig. 9.12. The wall thickness of 610 mm (24-in) diameter RCP is 90 mm (0.30 ft).

solution

1. Calculate the invert elevation at the critical point.

$$100 + 0.004\ [(\text{STA 8+78}) - (\text{STA 8+20})] = 100.23$$

2. Add the pipe diameter and wall thickness to get the top-of-pipe elevation.

$$100.23 + 610\ \text{mm} + 90\ \text{mm} = 100.93$$

3. Subtract the top-of-pipe elevation from the surface elevation.

$$101.26 - 100.93 = 0.33 \text{ m of cover}$$

Example 9.5   Calculate the amount of clearance over the top of the water main at STA 10+23 in Fig. 9.12. The wall thickness of 610 mm diameter RCP is 90 mm.

solution

1. Calculate the invert elevation at the critical point.

$$100.96 - 0.004 \ [(\text{STA } 10\!+\!60) - (\text{STA } 10\!+\!23)] = 100.81$$

2. Subtract the wall thickness.

$$100.81 - .09 \text{ m} = 100.72$$

3. Subtract the top-of-pipe elevation.

$$100.72 - 100.24 = 0.48 \text{ m of clearance}$$

When the responsible agency calls for criteria involving the hydraulic grade line or top-of-curb elevations, that information can be entered into additional columns of the storm drainage calculations form.

The calculation of the hydraulic grade line is required under some circumstances. The reason is that where the storm drain outfall is into a channel or stream, during heavy storms the water surface, which is the hydraulic grade line of an open channel, may be above the outfall elevation. In that situation, flow from the stream could back up into the streets from the stream. When that is a possibility, installation of flap gates to shut off flow between the stream and the storm drain system may be indicated.

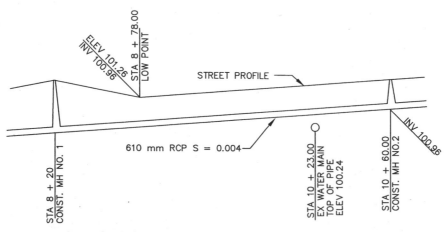

Figure 9.18   Storm drain clearance.

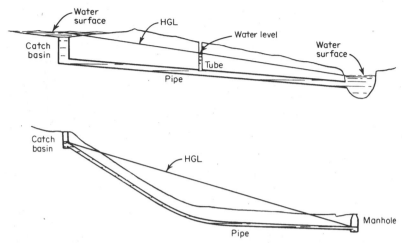

Figure 9.19 Hydraulic grade line (HGL).

The hydraulic grade line (HGL) is the water surface profile in an open system. In a closed system, pipes could be under pressure if the system is flat and shallow. The losses due to friction and junction boxes may become a factor forcing the flow upward into the street. When the pipe profile is in a sag vertical curve (Fig. 9.20), the HGL extends from the inside top of the pipe in the manhole or storm water inlet at one end of the pipe to the inside top of the pipe in the manhole or inlet at the other end and could cause a backup of the flow.

Technique for shallow sewers

When the site is nearly flat and the available outfalls are shallow, a different approach to designing the profile should be taken. When the horizontal layout is complete on the project master plan, the existing invert elevations should be marked in pencil. By starting at the outfall invert and applying the minimum allowable slope to the network of pipes, the sewers can be set at their deepest possible locations. The minimum slope is usually dictated by the agency and is based on the criterion that a minimum velocity of 0.6 to 0.76 m/s (2 to 2.5 fps) is necessary to assure sufficient scouring of the line.

Determine the invert elevations this way, and write them on the master plan in pencil. The size of the pipes is not important at this stage except that, when the connecting main is at minimum slope, the required diameter must not exceed the diameter of the outfall conduit. When the connecting pipe at minimum slope is larger than the outfall, its slope must be made steeper to accommodate the flow in the smaller pipe. The required diameter, in this case, is based on the total runoff Q_R that will be put into that outfall.

Estimate a value for C and a time of concentration t_c. Determine the required slope of the incoming pipe. Go through the same procedure with the sanitary sewer. Then calculate the clearances at crossings, and verify that there is enough clearance or enough flexibility in one of the systems to provide the clearance. It may be necessary to determine the pipe size based on the minimum slope at this time. If crossings cannot be made to clear, another layout should be tried. Existing utilities that are not dependent on gravity flow can be relocated to clear the new sewers. However, relocating existing lines should be avoided whenever possible.

When the site is too shallow to accommodate the storm system through the use of conventional methods, some alternatives that might be considered with respect to cost are to import fill to raise the site; to use small parallel sewer networks; to use an outfall that is deep enough though it is not adjacent to the site; to provide a detention pond or a percolation pond. A *percolation pond* is a reservoir constructed to hold water until it can seep into the ground. It is usually used to recharge the groundwater reservoir to provide for future water needs and to protect against subsidence of the ground caused by excessive withdrawals of groundwater by wells. In this case, the percolation pond becomes the outfall.

Outfalls

When the outfall is a river, stream, or channel, there will be a flood control district or other public agency with jurisdiction over any discharges into it. Usually an encroachment or construction permit is required. The agency may have its own specifications and standard plans that must be adhered to. Discharge into a natural waterway may require meeting concerns of a fish and game department or the Environmental Protection Agency as well. Be aware that failure to get the appropriate permits from some obscure agency can shut down construction. Do not be afraid to ask a spokesperson for one agency if he or she knows of other agencies that might have jurisdiction.

When a site is an otherwise unimproved area, there may not be any existing outfalls available. When this is the case, a temporary outfall will sometimes be allowed by use of a bubbler (Fig. 9.20). Bubblers serve the reverse function of storm water inlets. An inlet collects water from the surface and puts it into an underground piping system. A *bubbler* is a storm water inlet that takes water from an underground piping system and puts it onto the surface. The bubbler is installed at the end of the storm drain system. Runoff reaching the end of the storm system fills the bubbler and overflows through the grate, thereby becoming overland flow again. The top-of-grate elevation of the bubbler must be below the invert of the next upstream manhole or storm water inlet. Water will remain in the system wherever the elevation is below the top-of-grate

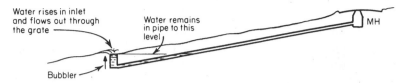

Figure 9.20 Bubbler.

elevation of the bubbler. Because of this, some jurisdictions will not allow bubblers. A dry well can be built around the bubbler and drain holes placed at the bottom in its sides for slow drainage of the line.

Small Individual Sites

Individual sites are graded to drain storm water overland to the street whenever possible. When this cannot be done, parking lots and landscaped areas are graded to create drainage basins. At low points, the water is collected in storm water inlets. The inlets are then connected to a piping network. If the lot is small, some agencies will allow a 76 mm (3-in) pipe or a set of 76 mm (3-in) pipes to be connected to the inlet and routed through the curb into the gutter, rather than into a piping system. Of course, there must be enough fall to provide sufficient slope, and the basin must be small enough for a 76 mm (3-in) pipe or pipes to have sufficient capacity.

Dry wells can be used to drain small basins or localized ponding where connection to an underground system is impractical. A *dry well* is a hole filled with gravel or other permeable rock. The size of the hole and the rock and the amount of space between the rocks and the amount of rock determines the capacity. The runoff can be stored beneath the surface in the interstices of the gravel while it percolates into the ground. The rate of percolation depends on the surrounding soil. If it is in a landscaped area, the rock may be left exposed or it may be overlaid with lawn. If it is in a paved area, a storm water inlet will be installed at the low point, and openings will be made to allow drainage from the inlet to the rock. The inlet and drainage openings must be deep enough so the drainage will not impact the integrity of the pavement. If the groundwater table is high because of a layer of impermeable soil, the well must be deep enough to reach below the impermeable layer.

Subterranean Water

Subterranean water can sometimes be a problem. The soils report may address this, but often the problem first becomes apparent during construction. Excessive groundwater can make construction of the improvements impossible, cause damage to building foundations, and greatly shorten the life of paved areas.

The underground ditch

One way of dealing with subterranean water is by constructing an underground ditch or french drain (see Fig. 6.1). A trench is dug to sufficient depth to protect the improvements from the flow of underground water. This task might entail encircling the improvements or simply constructing a drainage trench across a small, seasonal underground stream. A layer of permeable rock is laid in the bottom of the trench, and perforated pipe is installed with the holes down and at a slope that will allow the water to flow to some underground drainage system. The trench is then filled with permeable rock. The subterranean water flows into the trench, is drawn into the pipe, and is carried off. The perforated pipe must be connected to a storm drain system or other outfall.

High water table

A high water table can cause breakup of a roadbed in a short time. The problem arises as a result of traffic loads. As traffic moves over the roadbed, a pumping action is created. The saturated subgrade is pumped into the base rock section of the road, destroying its continuity and structural qualities. There are two possible solutions to this problem. One is the installation of french drains on either side of the roadway to draw down the water table (Fig. 9.21). The spacing and depth of these french drains should result from studies made by the soils engineer.

The other solution is to use a fabric membrane which can be wrapped around the entire base rock layer, separating it from the subgrade and subterranean water. This will ensure its continuity.

Non-Point-Source Pollution Control

Urban runoff from existing developments has been identified as a source of pollution and sediments in streams, wetlands, rivers, lakes, and bays. The pollutants of concern are oil, grease, and the toxics and heavy metals found in fertilizers and pesticides. Some jurisdictions

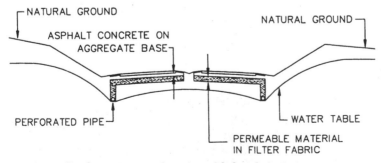

Figure 9.21   Roadway structural section with french drains.

now require site planning and pollution removal controls. These controls can be any of the following:

Reduction of the impervious areas to lessen runoff

Construction of detention ponds for silt removal

Construction of vegetated swales and trenches filled with pervious material to remove silt and promote infiltration

Construction of sand filter inlets similar to small-scale septic tanks to remove pollutants

Construction of wetlands

In the years ahead, environmental protection and enhancement will be a major issue worldwide. Civil engineers will be called upon to design safeguards for our environment.

Summary

Determining rainfall runoff for areas less than 120 hectares (320 acres) is accomplished using the rational formula. Ditches, culverts, and storm drain systems are used to carry the runoff safely away from people and structures. Manning's equation as applied in Chap. 8 and an analytical technique is used to determine pipe sizes. Underground water and the potential for water pollution must be addressed when developing property.

Problems

1. What is rainfall intensity?

2. What is meant by the term return period?

3. What does duration mean when used to determine runoff?

4. Why does rainfall intensity diminish over time?

5. Define C in the rational formula.

6. What is the runoff coefficient for a 4 ha (10-ac) site where 1.6 ha (4 ac) are building and paving, 2.4 ha (6 ac) are in landscaping of heavy soil with 1.2 ha (3 ac) at 1 percent slope and 1.2 ha (3 ac) at 10 percent slope?

7. Delineate the drainage basin contributing to point B on Fig. 9.3.

8. What is the volume of runoff at point B for a 100-year event from the basin delineated in problem 7? The basin is an unimproved area. Use Fig. 9.1 and Table 9.1.

9. How is grading used for storm drainage control?

10. What is the hydraulic radius for a trapezoidal ditch flowing full that measures 1 m (3 ft) across the top, 0.6 m (2 ft) across the bottom, and is 0.46 m (18 in) deep?

11. The ditch in problem 10 is made of concrete mortar and has a slope of 5 percent. What is its capacity?

12. Using the following information, what is the time of concentration at MH no. 3?

From MH no. 1 to MH no. 2:

  $t_c = 35$ min at MH no. 1

  $Q = 0.25$ m³/s (9 cfs)

  $S = 0.01$

  $L = 104$ m (340 ft); 460 mm (18 in) RCP

From MH no. 2 to MH no. 3:

  $Q = 0.34$ m³/s (2 cfs)

  $S = 0.04$

  $L = 55$ m (180 ft); 610 mm (24 in) RCP

13. A 200 mm (8-in) VCP crosses a 610 mm (24-in) RCP storm drain 46 m (152 ft) upstream from the storm drain manhole. The invert of the RCP at the manhole is 134.6. The invert of the VCP at the crossing is 145. What is the clearance?

14. What is a bubbler?

15. What is a dry well?

Further Reading

Brater, E. F., and Horace King, *Handbook of Hydraulics,* McGraw-Hill, New York, 1976.
California Division of Highways, *California Culvert Practices,* 2d ed., Sacramento, California.
County of Santa Clara, *Drainage Manual,* Santa Clara, California, 1966.
Dewberry & Davis, *Land Development Handbook,* McGraw-Hill, New York, 1996.
Federal Highway Administration (FHWA), *Hydraulic Design of Highway Culverts,* Hydraulic Design Series No. 5.
Haestad Methods Engineering Staff, *Computer Applications in Hydraulic Engineering,* Haestad Press, Waterbury, Connecticut, 1997.
Intelisolve, *The Hydraflow Library,* Marietta, Georgia, 1992.
Intergraph, *Storm Drainage Design,* Huntsville, Alabama, 1992.
Johnson, Frank L., and Fred F. M. Chang, *Drainage of Highway Pavements,* Highway Engineering Circular #12, Federal Highway Administration, McLean, Virginia, 1984.
Linsley, R. K., and J. B. Franzini, *Water Resources Engineering,* 4th ed., McGraw-Hill, New York, 1992.

Metcalf & Eddy, Inc., *Wastewater Engineering: Collection, Treatment, Disposal,* 2d ed., McGraw-Hill, New York, 1978.

National Technical Information Service (NTIS), *Drainage of Highway Pavements,* Highway Engineering Circular #12 (HEC-12), Springfield, Virginia.

National Technical Information Service (NTIS), *NTIS 1991 Catalogue of Products and Services,* U.S. Department of Commerce, Springfield, Virginia, 1991.

Reagan, Dwight, *Highway Drainage Design,* State of California, Department of Transportation, Sacramento, California, 1991.

State of California, Department of Transportation, *Highway Design Manual,* 5th ed., Sacramento, California, 1995.

Viessman, Warren, Jr., T. E. Harbaugh, and J. W. Knapp, *Introduction to Hydrology,* 3d ed., Harper, New York, 1990.

Woodward-Clyde Consultants, draft of "Storm Water Quality Controls for the Developments in Santa Clara Valley and Alameda County: A Guide for Controlling Post-Development Runoff," Oakland, California, 1992.

Water Supply Lines

Of the various aspects of land development, the design of water supply lines may be the easiest. This is primarily because water supply lines are force mains (pressure lines). They are not dependent on gravity to flow, so they can be designed to go over or under other underground facilities without significant loss of water pressure or velocity. Their horizontal and vertical placement is thus dependent only on convenience, protection from crushing if within a traveled way, and protection from freezing in cold climates. Though this chapter addresses water supply lines, the information generally holds true for other types of force mains as well.

Water Demand

The size of the demand for water depends on climate, distribution of land use, cost of water, availability of private sources of water, and cultural attitudes. This variability was illustrated earlier, in Table 8.1. In that table, the consumption rate is shown as 50 gallons per day per capita (gpd/cap) in Little Rock, Arkansas, and 410 gpd/cap in Las Vegas, Nevada. The demand also varies throughout the day and throughout the year (Fig. 10.1).

Industrial

Some industries have a high demand for water. If you are designing a water supply system from a public water supply for an industrial park, this must be taken into consideration. If the user industries are known, obtain expected consumption rates from them. The water consumption rates for some industries are shown in Table 10.1.

Nonindustrial

Consumption rates for various nonindustrial establishments are shown in Table 10.2. An increasing concern for water conservation may

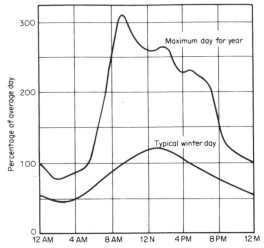

Figure 10.1 Hourly variation in water consumption at Palo Alto, California. (*From Ray K. Lindsay and Joseph B Franzini,* Water-Resources Engineering, *McGraw-Hill, New York, 1976, p. 430.*)

result in appliances which use less water than indicated here. Obtain from the architect expected water demand when designing water lines for high-demand buildings such as hotels and apartment houses.

A master plan which takes into account expected future growth should be available at the water supply agency. That agency may pro-

TABLE 10.1   Water Consumption in Representative Industries

Process	Consumption
Cannery:	
Green beans, gal/ton	20 000
Peaches and pears, gal/ton	5 300
Other fruits and vegetables, gal/ton	2 000–10 000
Chemical industries:	
Ammonia, gal/ton	37 500
Carbon dioxide, gal/ton	24 500
Gasoline, gal/1000 gal	7 000–34 000
Lactose, gal/ton	235 000
Sulfur, gal/ton	3 000
Food and beverage industries:	
Beer, gal/1000 gal	15 000
Bread, gal/ton	600–1 200
Meat packing, gal/ton live weight	5 000
Milk products, gal/ton	4 000–5 000
Whiskey, gal/1000 gal	80 000
Pulp and paper:	
Pulp, gal/ton	82 000–230 000
Paper, gal/ton	47 000
Textiles:	
Bleaching, gal/ton cotton	72 000–96 000
Dyeing, gal/ton cotton	9 500–19 000

SOURCE: Metcalf & Eddy, Inc., *Wastewater Engineering: Collection, Treatment, Disposal,* McGraw Hill, New York, 1972, p. 32.

TABLE 10.2 Estimated Water Consumption at Different Types of Establishments

Type of establishment	Flow (gpd/person or unit)
Dwelling units, residential	
Private dwellings on individual wells or metered supply	50–75
Apartment houses on individual wells	75–100
Private dwellings on public water supply, unmetered	100–200
Apartment houses on public water supply, unmetered	100–200
Subdivision dwelling on individual well, or metered supply, per bedroom	150
Subdivision dwelling on public water supply, unmetered, per bedroom	200
Dwelling units, treatment	
Hotels	50–100
Boarding houses	50
Lodging houses and tourist homes	40
Motels, without kitchens, per unit	100–150
Camps	
Pioneer type	25
Children's, central toilet and bath	40–50
Day, no meals	15
Luxury, private bath	75–100
Labor	35–50
Trailer with private toilet and bath, per unit (2½ persons)*	125–150
Restaurants (including toilet):	
Average	7–10
Kitchen wastes only	2½–3
Short order	4
Short order, paper service	1–2
Bars and cocktail lounges	2
Average type, per seat	35
Average type, 24-h, per seat	50
Tavern, per seat	20
Service area, per counter seat (toll road)	350
Service area, per table seat (toll road)	150
Institutions	
Average type	75–125
Hospitals	150–250
Schools	
Day, with cafeteria or lunch room	10–15
Day, with cafeteria and showers	15–20
Boarding	75
Theaters	
Indoor, per seat, two showings per day	3
Outdoor, including food stand, per car (3½ persons)	3–5
Automobile service stations	
Per vehicle served	10
Per set of pumps	500
Stores	
First 25-ft frontage	450
Each additional 25-ft frontage	400
Country clubs	
Resident type	100
Transient type, serving meals	17–25
Offices	10–15
Factories, sanitary wastes, per shift	15–35
Self-service laundry, per machine	250–500
Bowling alleys, per alley	200

TABLE 10.2   Estimated Water Consumption at Different Types of Establishments (*Continued*)

Type of establishment	Flow (gpd/ person or unit)
Swimming pools and beaches, toilet and shower	10–15
Picnic parks, with flush toilets	5–10
Fairgrounds (based on daily attendance)	1
Assembly halls, per seat	2
Airport, per passenger	2½

* Add 125 gal per trailer space for lawn sprinkling, car washing, leakage, etc.

NOTE: Water under pressure, flush toilets, and wash basins are included, unless otherwise indicated. These figures are offered as a guide; they should not be used blindly. Add for any continuous flows and industrial usages. Figures are flows per capita per day, unless otherwise stated.

SOURCE: Metcalf & Eddy, Inc., *Wastewater Engineering: Collection, Treatment, Disposal,* McGraw-Hill, New York, 1972, pp. 29–30.

vide and install all the water supply lines, or it may be necessary for construction plans of the water supply lines to be provided with other improvements. In either case, the water agency will dictate the size and types of water lines required.

Fire protection

A major factor in determining pipe sizes and water pressures is provision for fire protection. A minimum of a 150 mm (6-in) water line main should be used to serve residential areas. In areas with multistory buildings, 200 or 250 mm (8- or 10-in) mains may be required. In some cases, the domestic water source, the source for building sprinkler systems, and the source for the fire hydrants are all separate. When they are combined, fire protection demands draw from the other sources in an emergency.

On sites with commercial, industrial, or multifamily buildings, interior sprinkler systems are required for fire protection. A water protection assembly (WPA) (Fig. 10.2) is required. The responsible jurisdiction may have standard plans and specifications and will direct where the water protection assembly should be installed and space will have to be provided. The size of the assembly will be dependent on the size of the system. The purpose of the assembly is to provide easy access to the fire protection agent to see that the fire protection system is operational.

The Piping Network

The water company or agency may dictate the size, location, and types of pipes to be used. The sizing is based on fire protection requirements and expected consumption. There may be a standard location for water lines, such as 2 m (6 ft) from the face of the curb or 0.67 m (2 ft) behind the property line. Water lines that will be supplying drinking water

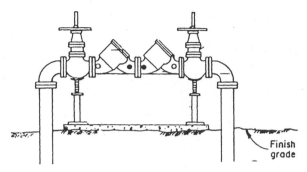

Figure 10.2   Water protection assembly.

must be protected from infiltration. This is partially accomplished by maintaining water pressure within the system.

For one- and two-story residential buildings in flat terrain, 45 pounds per square inch (psi) pressure is minimum. For water supply lines in steep terrain, 60 to 75 psi is recommended throughout the system. The pressure in water supply lines is usually provided by locating the water source on a hill or in tanks above the system, but water pressure can also be supplied by a pumping system. If the project site is at an elevation above surrounding terrain, check if the pressure will be sufficient for the project.

Water supply lines should never be placed in a joint trench with storm drain and sanitary sewer lines. A minimum horizontal distance of 3 m (10 ft) between water lines and sewer lines should be maintained. Pipes designed for high tensile strength as well as for crushing strength must be used for pressure mains. Cast-iron, cement-lined steel, steel, plastic, and, for large lines, reinforced-concrete pipes are all manufactured for pressure systems.

Horizontal layout

Water supply lines are laid out in a gridlike manner so that there are no unconnected ends. This is called *looping*. The exception to this is in short cul-de-sacs. The advantage of laying out the lines this way is that there are no dead ends in which water can stagnate and, if repairs are required, smaller areas will have their water supply cut off. Excessive demand put on the system by fire-fighting equipment causes less head loss (pressure loss) in a gridlike system. Sizing of the piping system is usually done by the water company using the Hardy-Cross solution or the nodal method of successive approximations. These techniques require successive iterations. Computer software programs are available to perform the necessary calculations. A description of these methods can be found in Brater and King's *Handbook of Hydraulics* (see "Further Reading" at the end of this chapter).

Where the water pipe is to curve, deflections specified by the pipe manufacturer should be used. Where there are to be connections, tees and wyes should be used and described on the plans. Where there is to be a change of direction either horizontally or vertically, a bend should be used and described. Most pipe manufacturers supply bends and elbows of 90°, 45°, 30°, 22½°, and 11¼° (Fig. 10.3). This limitation of available degrees of deflection for bends reduces flexibility of design.

Class equals crushing strength, lbs/lf (ASTM Test Method)	Pipe size Inside diameters, inches
1500	4, 5, 6
2400	4, 5, 6
3300	4, 5, 6

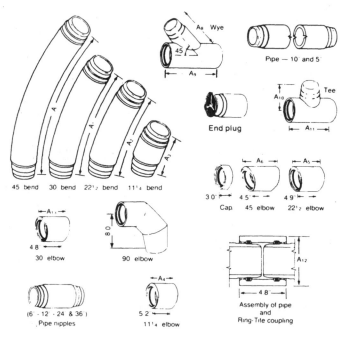

Dimensions, in

Pipe size	A	A₁	A₂	A₃	A₄	A₅	A₆	A₈	A₉	A₁₀	A₁₁	A₁₂	A₁₃
4	28.3	18.9	14.1	7.1	6.5	7.4	10.0	10.0	14.0	7.0	11.3	5.9	8.3
5	37.7	25.1	18.9	9.4	6.7	7.9	10.9	11.4	16.0	7.5	12.3	7.0	8.8
6	37.7	25.1	18.9	9.4	6.9	8.3	11.7	12.6	17.3	8.0	13.3	8.0	9.4

Figure 10.3   Bends and elbows.

The profile

Available elbows and bends must be used to make necessary changes of direction in the profile as well. Wherever the water line crosses other underground lines, a profile should be drawn and clearances verified. Inverts and slopes are calculated as they are for sewer lines; however, centers of pipes are matched rather than overts. Reducers, wyes, and tees are used for connections.

A profile can have a positive or a negative slope, but level sections should be limited to short distances. A slope of 0.005 is the minimum desirable, so that the lines will drain when maintenance is required. Wherever the profile changes from positive to negative and at the ends of lines, an air-relief valve should be provided. Bubbles of air trapped in the system will float to a high point. This accumulation of air can block the passage of water and stop the system from working (Fig. 10.4). An air-relief valve allows trapped air to escape.

Where the profile changes from negative to positive, a drain or blow-off valve may be required so that the pipe can be drained for inspection and repair. This valve can also be used to flush the line to remove rust and debris. The change of slope for each type of bend is tabulated in Table 10.3. Add the change of slope algebraically to the slope preceding the bend to determine the slope following the bend.

Example 10.1   Calculate the slopes for each of the sections of water line shown in Fig. 10.5. Use Table 10.3.

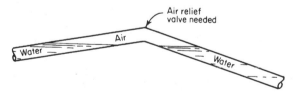

Figure 10.4 Bubbles of air trapped in a piping system will float to a high point and can block flow.

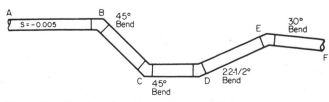

Figure 10.5 Section of piping for Example 10.1.

TABLE 10.3   Bends

	Degree of deflection	Change of slope		Degree of deflection	Change of slope
	90°	+2.0000		90°	−2.0000
	45°	+1.0000		45°	−1.0000
	30°	+0.5774		30°	−0.5774
	22½°	+0.4142		22½°	−0.4142
	11¼°	+0.1989		11¼°	−0.1989

solution

1. Calculate the algebraic sum of the slope *A-B* and the change of slope for a 45° bend.

−0.005	Slope *A-B*
−1.000	45° bend
−1.005	Slope *B-C*

2. Continue in the same manner for the other sections.

−1.005	Slope *B-C*
+1.000	45° bend
−0.005	Slope *C-D*

−0.005	Slope *C-D*
+0.414	22½° bend
+0.409	Slope *D-E*

+0.409	Slope *D-E*
−0.577	30° bend
−0.168	Slope *E-F*

When drawing the water line profile, remember that the horizontal and vertical scales are different, so a 45° bend will not result in a 45° angle on the profile. Because the forced system must be designed to circumvent the gravity systems, the space for elbows, bends, wyes, and tees may be limited. If this is the case, it is important to obtain the manufacturer's specifications. To verify the clearances, make a drawing using a natural scale.

Appurtenances

A variety of valves and other appurtenances are associated with force main systems. The purpose of air-relief, blow-off, and drain valves has been described. Some other valves you may encounter are check valves, pressure-relief valves, air-inlet valves, and pressure-regulating valves. Gate valves are located at many places in the system and are used to shut off the flow when there has been a break in the line. The system will also include fire hydrants, thrust blocks, and water meters, which are described briefly in this section. The various valves, bends, and tees should be shown on the plans and profiles (Fig. 10.6).

Gate valves.   The need to stop the flow of water at various locations in the piping network is met with gate valves. These valves are placed on each side of intersections and should be located no more than 400 m (¼ mi) apart along the water line. Their placement should be based on cutting off water to the least number of people when repairs or alterations are made. When water lines are located in streets, they should be placed out of the traveled way as much as possible. Each valve is

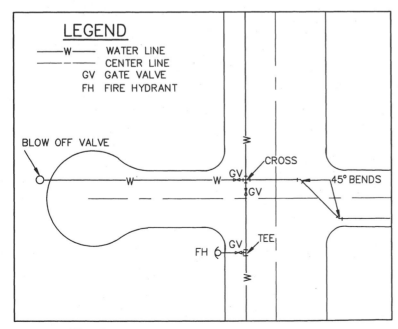

Figure 10.6   Plan showing water valves and bends.

enclosed in a valve box with a cover. The cover is usually marked with the word *water* and is made to withstand traffic loads when it is located in the street. In most cases, the valve box has a diameter of 150 to 200 mm (6 to 8 in), but large gate valves may have to be placed in manholes. An example of a gate valve is shown in Fig. 10.7.

Fire hydrants.   The locations of fire hydrants will be dictated by the fire department or another agency charged with public safety. Their type and placement will be based on what equipment is available to the fire department. Typically there is one at every intersection, so that hoses can be pulled in any direction. They should be placed no more than 150 m (500 ft) apart, to avoid excessive head loss in an emergency. A breakaway spool at the groundline will prevent damage below ground in case of an accident. A gate valve should be installed where the fire hydrant lateral connects to the main (Fig. 10.8). In some areas, the fire hydrant locations are marked in the street with blue traffic markers on the pavement, to facilitate location by fire crews.

Thrust blocks.   A change in the velocity or direction of flow in a pressure system causes a change in the direction and magnitude of the momentum (force = mass × velocity). To absorb the force of momentum and to anchor bends and tees, concrete thrust blocks are constructed (Fig. 10.9). The location and sizes of these thrust blocks will be described in standard plans and specifications or should be shown on a detail sheet. Where clearances for the force main must be checked, be sure to include the thrust blocks. The jurisdiction may provide standard plans.

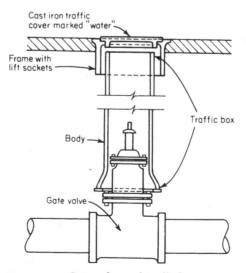

Figure 10.7   Gate valve and traffic box.

Figure 10.8 Fire hydrant with gate valve.

Laterals and meters. Water supply laterals to single-family homes are usually 20 to 30 mm (¾- to 1¼-in) pipes. The connections are made with a tapping device called a *corporation cock*. Flexible copper tubing or plastic pipe carries the water to the water meter. Connections for services larger than 50 mm (2 in) are made with tees and galvanized steel or plastic pipes. The type of water meter and meter box should be specified by the water supply agency. Each lot may have a lateral, or there may be one lateral with a connection to two meters.

Summary

Water supply lines are pressure systems; that is, they are not dependent on gravity for flow. As a result, the network and its coordination with other underground utilities is easier to provide. Providing adequate fire protection capacity, however, is a major consideration. The horizontal layout is different from sewer systems in that it is necessary to provide looping in the system wherever possible. The other significant difference is that it is a closed system, and wherever there is a change of direction, thrust blocks must be provided to accept the momentum from the force of the water.

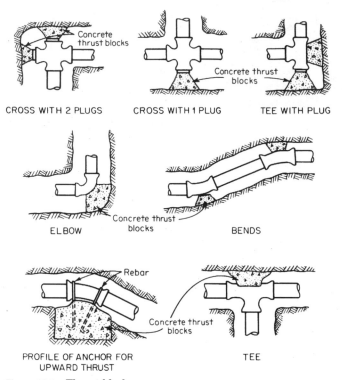

Figure 10.9 Thrust blocks.

Problems

1. Why should water supply and sewer lines be in separate trenches?

2. Why is looping used in water supply networks?

3. What is a WPA? What is its purpose?

4. What is the purpose of a gate valve?

5. Where should gate valves be located?

6. What is the purpose of thrust blocks?

Further Reading

Brater, Ernest F., and Horace King, *Handbook of Hydraulics,* McGraw-Hill, New York, 1976.

Haestad Methods, Cybernet, version 2, Waterbury, Connecticut, 1992.

Linsley, Ray K., and Joseph B. Franzini, *Water Resources Engineering,* 4th ed., McGraw-Hill, New York, 1992.

Viessman, Warren, Jr., Terrence E. Harbaugh, and John Knapp, *Introduction to Hydrology,* 3d ed., Harper, New York, 1990.

The Finished Plans, Specifications, and Estimates

The culmination of the engineering design is the preparation of the finished plans, specifications, and estimates (PS&E). They should be a clear, accurate representation of the completed engineering design and the construction work to be done. The plans will represent your company and its engineering expertise. If the drawings are sloppy, hard to read, or amateurish, people viewing the plans may expect that the engineering will be sloppy and amateurish as well.

Managing CADD Files

Now that you have learned the basics of the day-to-day engineering concepts and calculations, you will want to work with CADD to translate that work to a set of plans and specifications of which you can be proud. You will want to accomplish this in as little time as is reasonable, considering the scope of the task. Today, the key to success is closely linked to using CADD efficiently.

"To error is human. To really screw things up takes a computer" is a saying that was popular when computers first came into widespread use for business purposes. Now that CADD is in widespread use for our work, it is true for us. The use of CADD systems and the expansive availability and diversity of computer software is such that projects could become so screwed up that there is little choice but to trash some of the work and start over. We need to take steps to protect ourselves from having that happen.

It is imperative that engineers have a carefully thought-out plan of the organization of computer files. This is apparent for large engineering organizations but it is also true for individual consultants. It is too easy to give the same type of information similar but different names and then forget what information is contained in what file or layer name and have to check each of the similarly named files to retrieve what is needed.

How each company arranges its files cannot be dictated from an outside source such as this book. However, members of the company must make CADD file standardization a priority. The time spent will save hundreds of hours and I believe that those companies that have not made CADD-information organization a priority will not be able to compete with those who have.

Some issues about the use of computer information from allied sources is addressed in Chap. 5, Preliminary Engineering. If you have not already read the section in that chapter titled "Creating Preliminary CADD Drawings," it would be helpful to do that here. You may want to reread that section as a guide to establishing company policies on handling information received from others.

All the people in the company who will be using the CADD system from a just-hired junior engineer to the chief technical officer should be involved in deciding what standards will work best. It is especially important for your most accomplished CADD drafter to have input about choices for line types and fonts so the final products will have a sharp, clear appearance. If the company uses networking computers, if it has a research person, or if the Internet is being used, it may make sense to include the researcher and administration people as well.

There should be one person with the responsibility of maintaining the CADD standards and there should be someone assigned to take over in case the CADD manager is out of reach. The CADD manager may be a production engineer or someone whose sole purpose is managing CADD. It may be helpful to have a CADD consultant assist in creation of the standards, but the CADD manager must be a full-time employee. It is helpful to have a CADD consultant available on retainer or available on a call-for-fee basis. The time may come when this is not necessary, but in recent years it has almost always been.

Company Standards

Some of the things that should be standardized are listed here.

Job names

File names

Layers

Fonts

Line types

Colors

Blocks

Borders

X-references

Notes

Details

One way of organizing these things is with the use of trees (Fig. 11.1) similar to the way you outline a book or report—sections, chapters, headings, subheadings, and so on. Using this model, ask yourself what the main sections should be? Surveying, preliminary design, and final design is one set of sections that may work for your company. Or you may find it more useful to have the sections be public works, subdivisions for single-family residential, commercial, industrial, multifamily residential, parks, prisons, and schools. What you choose will depend on the character of the work most often performed by your firm.

General Standards

You may want to segregate General Standards into various categories such as Company Standards, Client Standards, Architects' Standards, and Public Works Standards. You may want to further divide different public works standards by different agencies—state highway, state general services, county, city no. 1, city no. 2, and so forth. General standards would include borders, symbols, fonts, line types, notes, and details. You may want to segregate notes and details into company, public works, and vendors.

Job File Standards

You may want to have a category for surveying projects which may or may not be associated with improvement plans. The surveying category for each project folder (directory) might have folders for each job or for each type of activity such as the following:

Control surveys

Boundary surveys

Topography

Aerial topography

ALTA maps

Tentative maps

Subdivision maps

Parcel maps

Final maps

Record of Survey Maps

Point maps

Lot line adjustment exhibits

FILE ORGANIZATION TREE

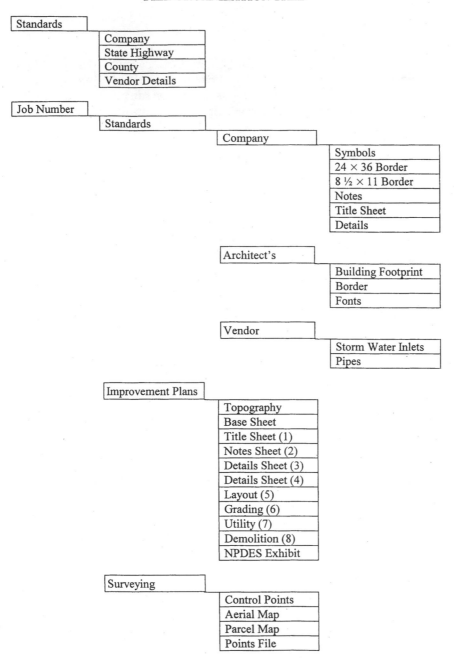

Figure 11.1 Computer file organization tree.

Improvement plans

If your company performs both public works and private developments, you may want to have separate categories for each. If your company does mostly public works projects, you may want to separate files by categories of public works with subsections of highways, street improvement, sewer systems, water systems, public buildings, and parks. You may want to use subsections for projects for private clients. The subsections might be subdivisions, commercial, industrial, single-family detached, and multifamily attached.

A word of caution! When deciding which categories to use, consider whether this will simplify your work or just add layers of information. Use categories only if by so doing, you will simplify your work and make your day-to-day operations more efficient.

Once these decisions have been made, ask yourself what subsections (files) should go into each job folder. You may want to have a section for job standards as taken from the general standards folder or you may find it more efficient to use the General Standards folder directly. If the project is to be constructed in phases, it may be helpful to have separate folders for the phases. One thing that might be helpful is to have a separate section for Resources from Others. That section might include subsections such as these:

Topography

Architect

Landscape architect

Mechanical engineer

Structural engineer

Lighting engineer

Fire safety engineer

Utilities designer (gas, electric, communications)

Vendors (details of prefabricated construction products)

If your company uses the computer to plot conceptual designs (preliminary engineering), the preliminary engineering and final plans should be in separate sections. This is true whether the project is a highway, a subdivision, or a site plan. By putting the conceptual plans into a separate folder, you reduce the risk of using the conceptual plan for calculations in the final phase of the work.

The preliminary folder will have the base drawings for the storm drain and sanitary sewer master plan, the base for the preliminary profiles for sewers and water lines, and the preliminary grading plan. The improvement plans section will have all of the plan sheets for the construction plans. Whether you designate them such as sheets 1 through 21 or such as cover sheet, details sheet, Edgehill Drive #1,

Edgehill Drive #2, George Road, Demolition Plan, Erosion Control Plan depends upon what your experience has been. The important thing is that you take time to plan how you want to handle the names of the files and then record them and make them company standards.

Naming the files

Once the decision has been made as to how to name the various files, the file names should be shortened for quick recognition and fewer key strokes. Even though many computer systems allow more, limiting the file names to eight characters is a good idea. Even fewer letters would be better. Try to use no more than two letters to describe each aspect of the file name. "IP01PS1.dwg" could stand for improvement plan, sheet 1, phase 1. The important thing is that the first two letters should be the part of the description looked for first when trying to find a file and that there be absolute consistency as to what each of the characters represents. In this case, the first two characters would always represent the type of plan, the second two always the sheet number, and any following characters supply additional information. Once the scheme has been decided upon, it must be described in an easy-to-access way for every engineer working on the plans. The scheme must be spelled out with the company standards policy and used consistently on all projects. Each file name on a project should be indexed with the definition of the acronym.

Layers

The number of layers on a project should be kept to a minimum. Company guidelines should be spelled out. The number of layers necessary will depend to some extent on the complexity of the project. The viewports and borders should have separate layers but beyond that, how the information is distributed among the layers should be decided using company guidelines. For a project with phases, you may want to create a different topography file for each phase by trimming away topography not needed for that phase. You may want to have a layer of surface topography and a layer for subsurface topography. It may be helpful to have a layer of existing topography and a layer of what the existing topography will be after demolition. Label the layers in a manner similar to the way the files are labeled. If your project will require many layers, keep in mind that you may want to filter some of the layers so that the name should begin with the letters that will allow you to select layers easily. One way of labeling a layer might be "EX-SD-PS1" for existing storm drain, phase one. You might be better served by using the storm drain designation as the first two letters resulting in such layers as:

SD-EX-PS1   For all existing storm drain information
SD-PS1   For line work

| SD-CB-PS1 | For catch basin and clean outs |
| SD-TXT-PS1 | For labels and text |

Most engineers have text on a different layer from the design line work. Catch basins and other facilities that will be a part of the surface drainage should be on a separate layer so that they can be incorporated with the grading drawing.

Line types and fonts

For company standards, your mantra should be "Keep it simple!" There seems to be a tendency among young engineers to play with all the wonderful options provided by the software so they may want to use many lettering styles, line types, and colors just because they can or because they think it makes them seem clever. But, in fact, it slows the work and slows their training as engineers. Always keep in mind that the purpose of engineering drawings is to convey information for construction of a project. Simplicity is elegance. "Keep it simple!"

If the client is a public works agency, it may dictate text styles, line types, and colors for display on the monitor. If not, two or three simple font styles should be enough for engineering drawings. A minimum-size lettering style may be dictated by the client or approving jurisdiction if the plans will be filed on microfilm. Also, a simple style will be easier to read if it is to be reduced on microfilm.

Select one style, size, and width for labeling all dimensions, grading elevations and slopes, pipe invert elevations, slopes, and lengths. For text of different sizes, as for street names, building names, pad elevations, or lot numbers, you may want to select different colors, thus line widths. You may want to select one other style such as a bold lettering style for the title sheet or street names. Border sheets might be the one place where different lettering styles can be justified for presenting such information as the company name and address and the client's name.

Keep the line types to a minimum as well. Surface improvements will all be with continuous lines. Lines representing existing conditions should be screened. There should be about three line widths for emphasis. You will probably need different line types for centerlines, property lines, easement lines, contour lines, possibly phase lines, and different line types for the different utilities—although lines with text serve well for utility lines. The more line types used, the more complicated the drawing will be. There is a point of diminishing returns for making more line types than are absolutely necessary. Beyond a certain point, having more line types makes the drawing more difficult to read. The more complicated the drawing is, the longer it will take to accomplish any drawing task.

At this time, engineering drawings are still being plotted in black and white. If your drawing is being created in more than sixteen colors (including screened line types), you are complicating your work unnecessarily. Avoid using black and colors close to black unless you *know* that all other consultants are using a white background screen. Otherwise, on those monitors where black is the background color, the black and near-black colors will not show up and so cannot be edited on screen, but they will plot.

"Keep it simple!" Leave the fancy fonts, line types, and colors to the marketing department and planners.

Setting Up the Drawings

The company should have a standard procedure for setting up drawings to facilitate the work. Time should be spent evaluating what size border will be used, how many sheets will probably be needed, and what scale to use. The more complex the project, the larger scale such as 1:200 (1 in = 20 ft) should be used. It is better to use one larger than needed than to use one too small. The trade-off is that if you have picked a scale too large, you will have more sheets in your plans than really necessary. If you select a scale too small, the plans will be difficult to label and assign notes. Further the plans will look cluttered, messy and be difficult to read.

The size of the border may be what the client dictates. If the client is a public agency, you will probably be using a border that it supplies. If the project is a large commercial, industrial, or multifamily site, the border size, if not the actual border, may be dictated. And the company should have standard border sheets of different sizes. Twenty-four by thirty-six has been a standard size for many years and there is no indication the sheet sizes will be changed to metric any time soon. Some agencies require a sheet size of 22 × 34 in because it can be reduced to half-size for easy handling and still be a usable scale. Further, when reduced to half-size, it will fold neatly into a report. This is handy for use with reports such as the NPDES report. 30 × 42 in sheets will probably work best for large commercial, industrial, and multifamily site plans.

Preliminary plans show concept, not design! It is too risky for preliminary plans to be used as a starting place in the computer for final improvement plans. The more revisions made to any drawing, the greater the potential for trouble. And the increased risk is not linear. The ramifications from a simple change can be widespread. After the preliminary engineering plans are complete, additional or different information may be dictated as a result of review by the jurisdiction. Ideally, when working on site plans, any questions of coordination with other consultants have been worked out before final design is begun. But, in spite of excellence among the consultants, there are always revisions as the work of each becomes more complete and more exact. Of course, the more complex and the more compact a site is, the more likely there will be changes. Making these changes and following through with the ramifications can be a daunting task. Further, information on the preliminary plan should have been shown to tenths of a meter (or foot) and will have to be expanded to show hundredths, so the additional time spent is not great. The client will not be sympathetic when delays result from errors caused by lack of perfect follow-through. The conceptual plans should be used as a guide not a base for the final plans.

A base sheet should be prepared. It will include all those elements which will be common to every design sheet. Typically this is street centerline, curb and gutter lines, and, if it is a site plan, building footprints.

Once the coordination among the grading plan and the storm and sanitary sewers has been carried out and the project master plan is complete, the master plan final design can be performed in the computer. The profile design of the sewers can now be taken from the preliminary profile work sheet and the final calculation of the profiles can be begun in the computer. Remember, the primary purpose of the improvement plans is to provide information for construction. The construction is performed from survey stakes and the survey stakes are located by the coordinates taken from your design. What this means is that you must not make any representations graphically. For instance, when showing a storm drain parallel with and 1.5 m north of the centerline, either use the offset command then change the offset line type to the needed line type, or better, go from a known coordinate point and traverse the desired line with distances and bearings. Then you can go back and ask the computer for distances after lines have been shortened or lengthened to fit the situation. The construction stakes will be located by the coordinates established this way.

When the final design master plan is complete, you will break it up into sheet-size drawings. Then open a new drawing and cross-reference in the other elements and labels. By waiting to label until the drawing (sheet) is otherwise complete, you will minimize relocation and possibly the rotation of information to make room for other information and give the plans a professional look. If you are using Autocad, be sure to do the final labeling in model space not paper space.

The California Department of Transportation dictates exact symbols and line and letter types and weights to be used on its plans, as well as which colors to use to illustrate the different line types on the electronic medium. Even with those controls, plans prepared by different engineers look different.

One of the most common mistakes is to show every elevation and every tree possible. This information may be helpful for engineering, but it may be of little importance for construction. If contours are used, are intermediate elevations really necessary? If it is necessary to show trees, must there be symbols for all of them, or can a note be added saying that those trees with trunks measuring less than 300 mm (12 in) in diameter are shown with a cross? Experienced contractors will not bid on a project without reading every note and examining the site themselves.

As previously stated, one of the problems with drafting is that too much material is presented. Even if the line work is carefully thought out, the plans can be a jumble. Too much information can be almost as bad as too little (Fig. 11.2). Topographic information should be cleaned up before construction drawings are prepared (Fig. 11.3). Turn off layers not required for construction.

The standard sheet

The approving agency or consulting firm may have specific requirements as to what must be included in a set of plans and how the plans are to be presented. If a checklist is available, obtain one and use it. (A checklist is included as a guide at the end of this chapter as Fig. 11.6.) All the sheets of a set of plans should have the following:

1. A border and a title block. The title block should show the project name, tract number, address, or whatever information will quickly identify it and distinguish it from other projects.

2. The name of the engineering consulting firm.

3. The signatures, license numbers, and dates of expiration of the licenses of responsible design engineers and approving agents.

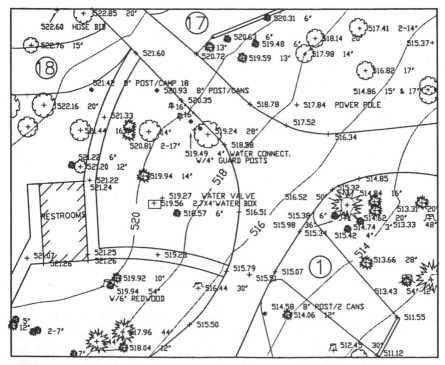

Figure 11.2 Topographic map with information needed for engineering but not construction.

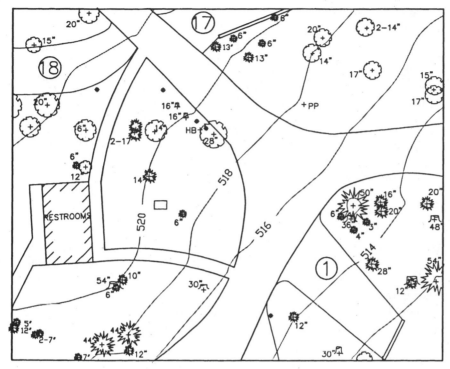

Figure 11.3 Figure 11.2 with extraneous information removed.

4. A sheet number together with the total number of sheets (for example, Sheet 1 of 10).

5. The scale or scales used.

6. The date of completion and approval of the plans.

7. A revision block with space to identify what revision was made, by whom, and on what date.

The standard sheet of specific size can have project specific information such as project names and the engineer's and developers names and addresses added to it and made a block. Then it can be moved to the project folder and inserted into the drawings as needed with ease and to provide consistency

When the plans are finished, prints are made and sent to the approving agencies for review. Agents examine and check the plans for completeness and compliance with their criteria. Changes and additions are marked on the plans or are listed. The amount of time this takes varies depending on the workload. When the changes and additions have been made, the revised plans are sent to the agencies for approval signatures. A set of the construction plans will be kept on hand at the public works department or approving agency. Some take the original drawings; others require reproducible copies. If the originals are to be relinquished, have good, reproducible copies made for your files.

The title sheet

Each set of plans should include a title sheet (see Fig. 4.6). What is appropriate to include on a cover sheet will vary from situation to situation. Some of the possibilities are described here.

1. A vicinity or location map should be included, showing where the site is located. This map may be at the center of the cover sheet and take up most of the space, or it may be placed in a corner and take up an area no more than 100 × 100 mm (4 × 4 in). The map should show major streets and have a north arrow, and the location of the site should be clearly delineated. It may or may not be drawn to scale.

2. A general plan showing the entire project to scale may be included to show the relative locations of streets and lots. The names of the roadways and, in subdivisions, the numbers of the lots, should be shown. Reference should be made on the general plan to which of the interior sheets of the plans shows the plan and profile of each street.

3. A legend should identify the meanings of the various symbols used throughout the plans and should be consistent for all projects where line types and symbols are not dictated by the client.

4. A basis of bearings used for the streets and lots should be included. An example of the basis of bearings is

> The bearing N 89° 30′ 08 E of the centerline of Main Street as shown upon that Record of Survey field in Book 24 of Maps, Page 8, Shasta County Records, was used as the basis of bearing.

5. The location and description of the benchmark should be shown. An example is

> An X is chiseled in the top of curb at the curb return on the northwest corner of Main Street at Second Street. The benchmark elevation is 103.36.

6. A table of contents or index of sheets is sometimes appropriate.

7. Construction notes may be included. For example, "All work must be done in accordance with the Standard Specification of the City of Springfield."

8. Various other details may be shown on the cover sheet.

Plan and profile sheets

The instructions on how to build the project are described with drawings on the plan and profile sheets (see Fig. 4.7) and with written instructions called *specifications*. Using these sheets, the surveyors will mark all the surface and subsurface improvements. Every point of change, whether a sanitary manhole or existing centerline, must be delineated and dimensioned so that the surveyors will be able to physically re-create the plans on the site. Surveyors locate significant points on the ground and mark those points with 2 × 2 wooden stakes called *hubs*. A simple description of what the hub represents is written on a 2 × 2 stake placed next to the hub. The various contractors then construct the improvements from the information on those stakes.

If a CADD system is being used, each aspect of the plan should be created on a different layer so there will be the flexibility to combine different elements as needed. For design of the project, it is necessary to have all the information on one sheet, the master plan, to coordinate the work, but the final plans should not include extraneous information. If CADD is not being used, the engineer must decide, based on the complexity of the project, how much information to include on each sheet. If the project is simple, it may be appropriate to show all of the surface and subsurface utilities on each sheet. At the other extreme, it may be best on a complicated project to show each utility on separate sheets.

On projects with a complex of buildings, such as apartments or schools, the work is typically shown on three types of sheets: one set for the layout, for locating buildings and streets; one set for grading; and one set for utilities. If there is to be significant demolition, this should be shown on a separate set of plan sheets.

The plan view.   Much of the information to be drawn on the plan views has been described throughout this book. Roadway names and a north arrow should be included on each drawing. For streets and roadways, the centerline of the roadway should be placed in the center of the plan-view space, oriented so that the north arrow will point toward either the top or the right edge of the sheet and the stationing reads from left to right whenever possible. The centerline is drawn with straight courses and circular or spiral curves. All the new construction will be tied to the centerline or reference line. The right-of-way and easement lines should be shown, and, on subdivision plans, the lot lines, lot numbers, and distances along lot frontages should be drawn along the right-of-way lines. For subdivision projects, it is useful to have the lot lines referenced to the centerline. This way, sewer and water laterals as well as driveways can be located and referenced.

Even though locations of utilities and other appurtenances will be located by their coordinates, they should also be described by their relation to the reference line. The layout of the storm and sanitary sewers is shown on the plan view. Manholes should be shown and numbered. This way, if the same manhole is shown on more than one sheet of the plans, it will not be counted more than once. The mains should be labeled with the length and type of pipe. Sanitary laterals, catch basins with their laterals, and water lines with laterals and water meters must all be shown. When there is an off-site utility line designed to go along an existing street or in an easement, it should be shown on the plan view and the profile. Information from the geometric cross sections should be shown at the ends of the sheets and wherever there is a change of dimensions. Dimensions to utility lines should be shown as well. Reference to the sheet where plans for adjacent or connecting roadways and utility lines are located should be placed at the connecting point, such as the roadway intersection. This may refer to another sheet on the same set of plans or, if the adjacent improvements are existing, to the tract or subdivision number.

When preparing the plans, particular attention must be directed toward the edges and boundaries of improvements. Wherever new streets connect with existing streets, a conform must be described and delineated on the plans. Where curbs are to meet, a note to "meet existing curb" and an elevation followed by "±" should be shown. New conduits will connect with existing lines or will enter existing manholes. Where conduits are to be connected directly to the end of an existing main, a note to "remove plug and connect" should be shown on the plan view. In some cases, it will be necessary to remove a flushing inlet or clean out at the end of an existing sewer line. Where existing water lines to be connected end in a blow-off valve, the valve will have to be removed. Where an existing manhole is to be entered with a new conduit, "break and enter" should be written on the plans at the manhole. Where sanitary laterals are to be connected to existing mains, reference to construction technique (for example, "tap and saddle") must be noted. It may be necessary to remove existing power poles and relocate other utility facilities. All these activities must be described so that their costs can be determined and construction progress can be properly coordinated.

The profiles. A grid is provided to show the profiles. It should be stationed, and elevations should be shown at the edges. This facilitates plotting and reading. The horizontal scale will match the plan view, but the vertical scale is usually different to better illustrate vertical information. Unless the slope is steep, a vertical scale should be selected that will show the full length of the profile without running off the grid vertically.

Show and identify roadway centerlines. Grades and vertical curve data must be shown. The stations and elevation of the beginnings and ends of vertical curves must be shown and identified. The lengths of the curves, the stations and elevations of the points of intersection, and the grades must be shown as well. It will be necessary to give elevations on the profile of vertical curves at the high and low points, at the midpoint, and at intervals frequent enough for surveyors to mark the profile for construction of a smooth roadway. This distance may be at quarter points or eighth points. Some even interval, such as every 10 m (25 ft), may work better.

Where slopes are steep, it may be necessary to break the profile at a match line and continue with a different elevation (Fig. 11.4). Where it has been necessary to use an equation in the plan view, that equation must be shown on the profile (Fig. 11.5). This is accomplished by restationing the profile from the next even grid line and continuing the stationing from there. The elevation must be the same on the stationing backward as on the stationing forward. Avoid locating any change of slope at an equation, as the match will be less apparent and may cause confusion. Elevation equations can happen, but are rare. The existing natural ground line at the centerline must be drawn and labeled. When the profile is not within a street right-of-way, show the proposed finished ground profile.

Pertinent underground facilities must be shown and identified. If there are too many existing and proposed utilities to show them all clearly on one profile, it may be necessary to provide separate areas of profiles. The existing and proposed utilities on one side of the street can be shown as one profile, and the information for the other side of the street can be shown on another profile. This approach should be used only in extreme cases, as conflicts and crossings will be less apparent and more likely to be missed.

At the ends of the profile will be an equation with a connecting street or the tract boundary. A note to "conform to existing" may be appropriate. The storm and sanitary profiles may be shown with a symbol, a single line, or a double line to indicate the top and bottom of the pipe. The sanitary sewer may be shown with a single solid line, and the storm sewer may be shown with a single dashed line. The manholes are drawn to scale or are shown symbolically with slender triangles. If there is any question of fit, show the manhole to scale either on the plan and profile

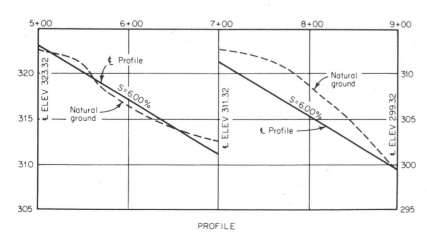

PROFILE

Figure 11.4 Steep slopes may require that the profile be broken.

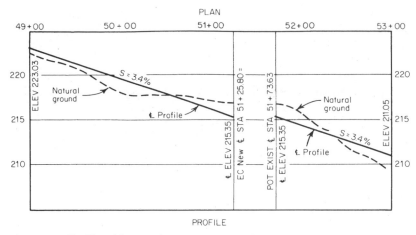

Figure 11.5   Profile with equation.

or in a detail. The manholes should be labeled, such as "Construct Sanitary M.H. #3 1.5 m (5 ft) lt. STA 3+23.56." The invert elevations should be labeled and identified, such as "380 mm (15 in) RCP INV 221.53 thru, 300 mm (12 in) RCP in S. INV 221.78." Elevations on street and utility profiles should be labeled at the sheet match line.

Drafting style can serve to clarify what might otherwise be confusing. If there are several profiles in the same vertical space, different symbols can be used for each, or double-line drawing can be used with hidden (dashed) lines representing places where one conduit is behind another. When the double-line approach is used, the symbol for a manhole should be removed where the conduit connects. Conduits behind manholes should be drawn as hidden lines. Existing conduits should be shown with lighter lines than proposed conduits.

When the drafting is complete, recalculate the profiles from the finished drawings, being careful to truncate elevation values beyond thousandths of a meter (hundredths of a foot) at grade breaks. This is a good check for drafting errors and ensures that values calculated will agree with those shown, regardless of where the profile calculation is begun. Finally, the checklist should be consulted to be certain that everything has been done (Fig. 11.6).

Detail drawings

It may be necessary to include sheets of detail drawings. Where there are no standard plans and where the situation requires unusual or unique structures for which no standard details are available, the structures must be designed and drawings prepared. If many details are necessary, it may be useful to designate sheets as, for instance, "Storm Drain Details" or "Water Facilities Details." A clear title, such as "Sanitary Manhole 1.5 m (5 ft) left of STA 50+20 on Main Street,"

```
┌─────────────────────────────────────────────────────────────────────────────┐
│                        A FINISHED PLANS CHECK LIST                            │
│                                                                               │
│  Cover Sheet                                                                  │
│          Title block                                                          │
│  _____ Name and address of project                                         │
│  _____ Name and address of approving agency                                │
│  _____   Names of approving agents with titles, stamps, signatures, and     │
│              license expiration dates                                         │
│  _____   Names of approving departments                                    │
│  _____ Name, address, and phone number of consulting engineer              │
│  _____ Consulting engineer's stamp, signature, and license expiration date  │
│  _____ Date                                                                 │
│  _____ Scale                                                               │
│  _____ Revision block w/date and initials                                  │
│  _____ Sheet number and total sheets                                       │
│          A general plan                                                       │
│  _____ Street names                                                        │
│  _____ Lots                                                                │
│  _____ Blocks                                                              │
│  _____ Adjacent tract numbers                                              │
│  _____ Scale                                                               │
│  _____ North arrow                                                         │
│          A vicinity map                                                       │
│  _____ Street names                                                        │
│  _____ The site                                                            │
│  _____ A north arrow                                                       │
│  _____ Scale                                                               │
│  _____ Index of sheet's plan                                               │
│  _____ Legend                                                              │
│  _____ Basis of bearings                                                   │
│  _____ Bench mark(s)                                                       │
│  _____ Details                                                             │
│  _____ Street structural cross section                                     │
│  _____ Notes                                                               │
│                                                                               │
│  Plan and Profile Sheets                                                      │
│          Title block                                                          │
│  _____ Name and address of project                                         │
│  _____ Name and address of approving agency                                │
│  _____   Names and titles of approving agents with stamp, signature, and    │
│              license expiration date                                         │
│  _____   Names of approving departments                                    │
│  _____ Name, address, and phone number of consulting engineer              │
│  _____ Consulting engineer's stamp, signature, and license expiration date  │
│  _____ Date                                                                 │
│  _____ Scale                                                               │
│  _____ Revision block w/date and initials                                  │
│  _____ Sheet number and total sheets                                       │
│  _____ Stationing                                                          │
│  _____ Elevations                                                          │
│  _____ Street name(s)                                                      │
│  _____ Name of reference line(s)                                           │
│  _____ North arrow(s)                                                      │
└─────────────────────────────────────────────────────────────────────────────┘
```

Figure 11.6 Finished plans checklist.

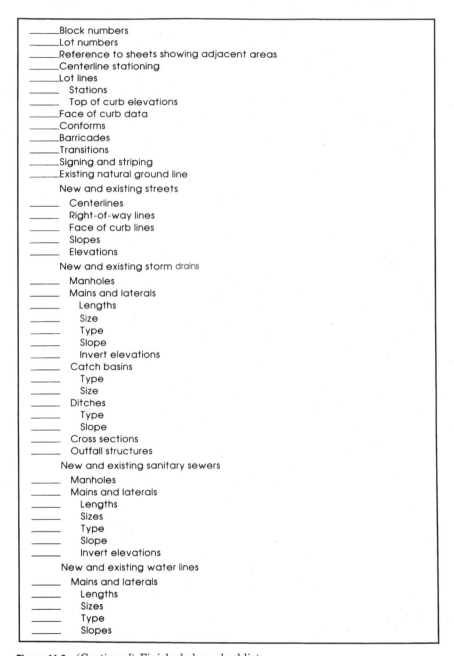

_____Block numbers
_____Lot numbers
_____Reference to sheets showing adjacent areas
_____Centerline stationing
_____Lot lines
_____ Stations
_____ Top of curb elevations
_____Face of curb data
_____Conforms
_____Barricades
_____Transitions
_____Signing and striping
_____Existing natural ground line
    New and existing streets
_____ Centerlines
_____ Right-of-way lines
_____ Face of curb lines
_____ Slopes
_____ Elevations
    New and existing storm drains
_____ Manholes
_____ Mains and laterals
_____   Lengths
_____   Size
_____   Type
_____   Slope
_____   Invert elevations
_____ Catch basins
_____   Type
_____   Size
_____ Ditches
_____   Type
_____   Slope
_____ Cross sections
_____ Outfall structures
    New and existing sanitary sewers
_____ Manholes
_____ Mains and laterals
_____   Lengths
_____   Sizes
_____   Type
_____   Slope
_____   Invert elevations
    New and existing water lines
_____ Mains and laterals
_____   Lengths
_____   Sizes
_____   Type
_____   Slopes

Figure 11.6   (*Continued*) Finished plans checklist.

should be provided. Drawings using different scales may be grouped together on the same sheet, but if this is the case, each drawing should be clearly labeled as to scale. If prefabricated structures such as drainage inlets are being used, you may be able to download them from the manufacturer's web site.

```
_____    Bends
_____    Elevations
_____    Water meters
_____    Fire hydrants
_____    Valves
          New and existing gas lines
_____    Size
_____    Type
_____    Valves
          New and existing electrical lines
_____    Size
_____    Type
_____    Power poles
_____    Transformers
          New and existing electroliers
_____    Conduit and conductors
_____    Junction boxes
          New and existing telephone lines
_____    Poles
_____    Junction boxes
_____    Manholes
          Notes
```

Figure 11.6   (*Continued*) Finished plans checklist.

Specifications

Specifications are written descriptions of materials and procedures to be used for construction of the project. They include such information as site safety requirements, what mixes should be used for concrete, methods of materials measurements, and payment schedules. States and large cities have standard specifications which are referred to when projects are within their purview. Prewritten specifications are available for various types of projects. These are sometimes referred to as master specifications. Paperback and computer software copies of these specifications are available.

The specifications are organized in sections—for example, Section 39 is Asphalt Concrete, and Subsection 39-303A(1a) is Manual Proportioning for Batch Mixing with Hot Feed Control of Asphalt Concrete. Engineers often refer to specific sections from state specifications rather than rewriting the needed instructions.

There is also a master list of titles and numbers for the construction industry in book form. Some clients or architects will require use of these titles and numbers to provide consistency. Most companies have specifications from previous jobs which can be revised for new jobs. But just using a generic set of specifications from another job is absolutely unacceptable. It is extremely important that when specifications are written for a particular job, they address the exact situation that the contractor can expect to encounter on that job. For instance, if the plans call for Type A Traffic Reflectors to be installed with epoxy and the method of preparing the surface for placement and the quality of the epoxy is not described, those traffic reflectors may be installed using what is cheapest and quickest and result in a traffic delineation product that does not last.

Estimates

Throughout the design process estimates have been made regarding the cost of construction of the project. When the plans are complete, exact measurement of construction materials can be made. The lengths of curbs, tons of asphalt, and all of the materials and activities needed to construct the project are now known and can be measured with certainty. These quantities are carefully measured and listed for inclusion in the bid packages.

The engineer performs this work and prepares a formal estimate of the cost of construction. Some developers want the engineer's estimate to be kept secret until after the bidding. Contractors will also make an estimate and determine what their costs of construction will be before they bid on the project. The engineer's and the contractor's estimates are compared. If they are very different, the reason should be understood. Perhaps a poor economy has forced contractors to bid low because they need the work. But contractors whose bids are very low may not have a good understanding of the work involved. If contractors' bids come in high, the developer may want to seek out more contractors to bid on the project.

Summary

Today the key to success is closely linked to using CADD efficiently. Taking the time to standardize procedures and elements is essential to remaining competitive. A perfectly engineered project is of little use if the plans for it are not clear. To accomplish this, expert draftsmanship is necessary.

Specifications are a detailed, written description of methods and materials to be used to construct the project. It is the engineers' responsibility to prepare the specifications. Engineers must also prepare a final, exact estimate of construction costs.

Problems

1. Why is standardization of CADD procedures, fonts, line types, and layers important?

2. How does the appearance of the plans affect the cost of construction?

3. What three questions must be answered before planning the drafting?

4. What is a standard sheet?

5. Name five things that may appear on the cover sheet.

6. Civil site plans for apartment complexes are usually shown on three types of sheets. What are they?

6. Edges of sheets are important. Why?

7. Show the centerline profile from STA 100+00 to STA 110+00 on grid paper. The elevation at STA 100+00 is 255. The slope is –5 percent. The following centerline equations must be shown.

Old Highway 1 STA 101+02 = New Highway 1 STA 101+86

Old Highway 1 STA 105+54 = New Highway 1 STA 105+32

8. What are specifications?

Further Reading

CAD Masters, Inc., Softdesk Civil Workshop, Walnut Creek, California, 1997

Construction Specifications Institute, Masterformat, Manual of Practice, Alexandria, Virginia, 1988.

Construction Specifications Institute, Spectext and Spectext II Hardcopy, Alexandria, Virginia, 1992.

Ruggeri-Jensen-Azar and Associates, CADD Guidelines, Milpitas, California, 1997

Sandis Humber Jones, CADD Standards Manual, Menlo Park, California, 1994.

State of California, Department of Transportation, Standard Specifications, Sacramento, California, 1988.

12

The Construction Phase

Once the plans are complete and approvals and permits have been acquired, the project moves into the construction phase. The design engineer must be available for clarification of the plans and to advise or redesign if required because unexpected problems occur during construction. However, it is unusual for civil site engineers to be actively involved in construction. It is the responsibility of the construction superintendent and inspectors from the approving jurisdiction to verify that the plans are followed.

Contracts

The developer will ordinarily prepare the contracts for construction, but the design engineer will be asked to prepare a bid package of quantities. It is desirable for engineers to have an understanding of the kinds of information that should be included in the construction contract. Some of the elements of the construction contract are listed here.

The number of days until commencement of the work

The number of days until completion of the work

Penalties levied for not meeting deadlines

Incentive bonuses for completion ahead of schedule

Definition of extra work

Methods of payment

Responsibility for inspection

Responsibility for survey stakes

Procedures for dealing with discrepancies

Performance bond requirements

Liability and worker's compensation insurance requirements

Requirements for providing equal opportunities for workers

Requirements for the condition and safety of the site

Requirements for dust and erosion control

The plans, specifications, and estimates (PS&Es) are part of the contract. The contract will include a list of the quantities and bid items formatted in a way to make the contractor's bid clear.

Quantities

An important part of the contract is a list of the items to be built. The contract may state that the list is an estimate and that the contractor is responsible for building the project regardless of what is required. What this means is that the contractor must calculate his or her own estimate of the material needed and be satisfied with the list provided. The exact length of the curb and gutter and the square meters (or footage) of paving needed can be calculated from the plans. Particular attention must be given to the descriptions when there is more than one type of the same kind of item. For example, adjusting to grade of a manhole rim that must be raised will cost less than adjusting to grade of a manhole rim that must be lowered.

Installing a sanitary sewer main before paving is laid costs less than installing a sewer main when existing pavement must be removed and replaced. Catch basins with galleries cost more than standard catch basins. Class IV RCP costs more than Class III RCP. Items dimensioned in linear meters (or feet) should be precise to within a 0.1 m (or 1 ft). Items dimensioned in square meters (or feet) should be precise to within 1 m^2 (or ft^2) if the total is less than 10 m^2 (or 100 ft^2). Earthwork quantities should be rounded to 5 m^3 (or 5 yd^3). It may be necessary to convert square meters (or feet) of paving or aggregate base to tonnes. The conversion factors depend on the asphalt mix and on the type of aggregate base being used, so these factors must be determined for each item.

Spreadsheets

Many developers—whether public or private—make use of a spreadsheet (Fig. 12.1). The spreadsheet lists the items and their quantities along one edge and the contractors who are bidding for the project along another edge. For each contractor, a unit price and a total price are listed across from each item. The prices for all the items for each contractor are totaled. This allows easy comparison of individual unit prices as well as totals. Great care should be taken to assure that the quantities are accurate. An unscrupulous contractor could charge for materials not used if the quantities are greater than needed. But you

		SPREADSHEET								
Quantity	Description of item	Engineer's estimate		Contractor		Contractor		Contractor	
		Unit price	Total	Unit price	Total	Unit price	Total	Unit price	Total
		Total		Total		Total		Total	

Figure 12.1 Spreadsheet.

can be sure that if the quantities are wrong but under-reported on the quantities list, the contractor will ask to be paid for them.

It is helpful to get a copy of these spreadsheets and keep it on file. A file of spreadsheets makes an ideal reference for estimating costs for similar projects. Further, it is a good check of your own work against that of someone else.

Construction

When the project goes into construction, your responsibility is not over. Any work you have done incorrectly or incompletely and any design or specification that is not clear will come back to haunt you. If you have been thorough in your research and single-minded in resolving any uncertainty, you can enter the construction phase with confidence.

Scheduling

On large construction contracts, construction schedules are prepared to ensure that the work will progress smoothly. It is unlikely that the design engineer will be involved in this activity, but he or she should at least be aware of the factors involved in scheduling. The method often used is simple bar charts with the number of days or weeks across the top and the activities listed down the left side. There is a bar for each activity to be accomplished. The beginning of the bar shows the time that activity will begin and the end of the bar shows the time that activity will end. These charts are helpful for seeing what activities will be taking place at any particular time and for reference to subcontractors.

Bar charts are limited in that they do not show the interrelations and interdependencies which control the progress of a project. A number of techniques can be used to illustrate that interdependency. The critical path method (CPM) is the most commonly used, so it is discussed here. Figure 12.2 is an example of a CPM chart. The chart is read from left to right. The progression of activities from 1 to 13 is shown in the circles in the lines. There are numbers in circles and squares off the lines which represent the total number of days to reach that point in the work where they are shown. The numbers in the circles represent the number of days, if there is an early start of that activity. The numbers in the squares indicate the total number of days to that point if the activity got a late start. The numbers on the lines indicate the number of days needed to accomplish the activity shown on the line. Lines without numbers indicate materials that must be delivered to the site before the indicated activity at the arrowhead can take place.

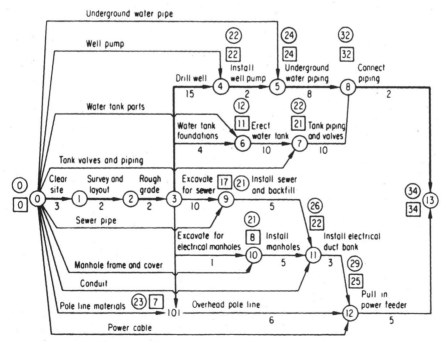

Figure 12.2 CPM chart. (*From James J. O'Brien,* Construction Management, *4th ed., McGraw-Hill, New York, 1992.*)

Some developers require a bar, CPM, or similar chart for projects they have built. The value for construction management seems apparent, and simply going through the exercise contributes to preparation for construction activities. In actual practice, the use of the chart is minimal after the work begins.

The Prebid and Preconstruction Meetings

Often prebid and preconstruction meetings are planned. Representatives of the developer, engineer, and contractors meet and discuss how the work will progress and to arrange coordination. Questions may arise about conditions on the site or details in the plans. Review the plans, specifications, and quantities list before attending, so that the information will be fresh in your mind. Be sure to take a set of plans and specifications with you. Do not rely on others to supply them. If you are asked a question you are not prepared to answer, offer to get the answer as soon as possible. Any question asked after the prebid meeting by one contractor may have to be addressed through the client and distributed to all potential contractors to be fair so that all will have the same information.

Construction Troubleshooting

Occasionally, unanticipated situations become apparent during construction. Ancient utility lines are uncovered, a vein of hard rock makes conventional trenching techniques impractical, or neighbors think the work is encroaching on their land. When this happens, the people involved will look to the engineer for explanations and direction. This is always an emotionally charged situation because delays are so expensive. It is imperative that your appearance be one of calm self-confidence. Solutions to the problems will be obscured when tempers flare and fears are exaggerated. Often it is difficult to get a clear understanding of the problem until people are calm.

Initially, there will be a phone call reporting that there is a problem and asking you to go to the site. Get as much information as possible. Ask the caller to be specific. By insisting on a clear explanation, you make the caller think the problem through. Sometimes this is all that is needed. The problem may be solved more quickly if you do *not* go to the site. You will have more resources in the form of maps, plans, computers, and other engineers in your office than in the field. In the process of explaining the problem in a clear way, the caller may realize the solution immediately or with a little additional information from you.

When it is necessary to go to the site, go prepared. Take along maps and plans that may be helpful. A plan with good preconstruction topography will be helpful. The cost of a construction stoppage can be extremely high. Get the work moving again as quickly as is reasonable.

However, workers' safety must be of paramount concern. Be open to suggestions from the workers on the site. They may be more familiar with the site conditions, and their experience can be a valuable asset. Acknowledge that you respect their experience and value their input.

When the problem is resolved, prepare a letter to interested parties and to the job file. Weeks or months later, you may not recall why the solution chosen was best. When changes are made during construction, it is important that they are marked on the as-built plans and that all affected parties are informed.

Summary

Design engineers' work is not complete until the project is constructed and the total engineering fee paid. Engineers must be available to advise the client about unusual circumstances that should be addressed in the contract, clarify any part of the plans that is not clear, and resolve any unexpected conflicts that arise during construction. Preparation of quantities for the contract bid package also falls upon the engineer. This is a simple task, as the work will have been done while preparing the estimate.

Problems

1. What contribution does the design engineer make to the contract?

2. What is CPM, and how is it used?

3. What are as-builts?

Further Reading

O'Brien, James J., *CPM in Construction Management,* 4th ed., McGraw-Hill, New York, 1992.

Conversion Table

Lengths

1 inch (in)	25.4 millimeter (mm)
39.37 inches	1 meter (m)
12 inches = 1 foot (ft)	
3 feet = 1 yard (yd)	
1 mile = 5280 feet	1.609 kilometers (km)
1 chain (surveyor's) = 100 feet	
1 link (surveyor's) = 1 foot	
1 chain (Gunter's) = 66 feet = 4 rods	
1 link (Gunter's) = 0.66 feet	
1 rod = 16.5 feet	

Area

1 square inch (in^2)	645 square millimeters (mm^2)
10.76 square feet (ft^2)	1 square meter (m^2)
1 square yard = 9 square feet	
1 acre = 43,560 square feet	0.4047 hectare (ha)
2.47 acres	100 are = 1 hectare
100,000 square (Gunter's) links = 160 square rods	
10 square (Gunter's) chain	4046.87 square meters
1 square (Gunter's) chain = 0.1 acre	
1 section = 1 square mile = 640 acres	
1 square mile = 27,848,400 square feet	

Volume

35.28 cubic feet (ft^3)	1 cubic meter
1 acre foot = 43,560 cubic feet	
1 U.S. wet gallon	0.833 imperial gallons = 3.785 liters
1 cubic foot = 7.48 gallons (gal)	
1 cubic yard = 27 cubic feet	

Discharges

1 cubic foot per second (cfs) = 449 gallons per minute (gpm) = 0.646 million gallons per day (million gpd or mgd)	0.0283 cubic meters per second (m^3/s)
35.3 cubic feet per second (cfs)	1 cubic meter per second

Weight

1 pound (lb)	454 grams (g)
1 U.S. ton = 2000 pounds	
1 long ton (Great Britain) = 2240 pounds	1016 kilograms (kg)
62.4 pounds per 1 cubic foot of water	
1 gal water = 8.33 pounds	
1 U.S. sack of cement = 94 pounds	
1 U.K. sack of cement	50 kilograms (kg)
1 Canadian sack of cement = 87.5 pounds	

Helpful Trigonometry

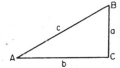

Sine (sin) of an angle = side opposite divided by hypotenuse

$$\sin A = \frac{a}{c}$$

$$\sin B = \frac{b}{c}$$

Cosine (cos) of an angle = side adjacent divided by hypotenuse

$$\cos A = \frac{b}{c}$$

$$\cos B = \frac{a}{c}$$

Tangent (tan) of an angle = side opposite divided by side adjacent

$$\tan A = \frac{a}{b}$$

$$\tan B = \frac{b}{a}$$

Law of sines:

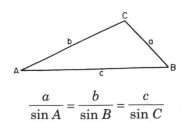

$$\frac{a}{\sin A} = \frac{b}{\sin B} = \frac{c}{\sin C}$$

Helpful Geometry

Circle

Circumference $= 2\pi r$

Area $\qquad = \pi r^2$

Sphere

Surface area $\quad = 4\pi r^2$

Volume $\qquad = \frac{4}{3}\pi r^3$

Ellipse

Area $= \pi ab$

Triangle

Area $= \frac{1}{2}bh$

Trapezoid

Area $= \dfrac{b_1 + b_2}{2}h$

Circular Cylinder

Volume $= \pi r^2 l$

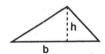

Right Circular Cone

Total surface area $= \pi r l + \pi r^2$

Volume $\qquad = \frac{1}{3}\left(\pi r^2 h\right)$

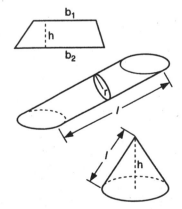

Glossary

AASHTO The American Association of State Highway and Transportation Officials, Washington, DC.

AB aggregate base.

A.B. anchor bolt.

ABM assumed benchmark.

AC asphalt concrete.

accurate correct though may not be precise.

ACEC American Consulting Engineers Council, 1015 15th Street, NW, Washington, DC 20005.

ACP asbestos cement pipe.

ADA Americans with Disabilities Act

adverse grade the condition in which a conduit slope is positive (up) and groundline profile over the conduit is negative (down).

adverse possession a method of acquiring ownership of property by occupation for a statutory period of time; ownership must be determined by a court of law.

ADT average daily traffic; the number of cars passing a particular point on an average day.

aerial relating to or occurring in the air; for example, aerial power lines are mounted on poles, and aerial photographs are taken from an airplane.

agency political body responsible for approval of a particular aspect of the design, such as city, county, flood control district, sanitation district.

aggregate clean, broken rock used for the preparation of concrete and asphaltic paving and as base material for structures; classified according to size between sand and 76 mm (3 in).

ALTA American Land Title Association.

anchor bolt a bolt with the threaded part projecting from the concrete to secure framing or other structural members.

angle of repose angle formed with level ground that a loose pile of material assumes naturally.

annexation legal procedure by which a political area such as a city can add more territory.

APN assessor's parcel number; number assigned to a parcel of land by the county assessor for property tax purposes. Reference number used to retrieve a sketch of the property and surrounding properties.

appraisal   determination of the monetary value of real property.

appraisal map   drawing showing property lines and certain topographic features for the purpose of identifying property for establishing monetary value.

appurtenance   addition or accessory to something else.

APWA   American Public Works Association, 106 West 11th, Kansas City, MO 64015.

AS   aggregate subbase.

ASB   aggregate subbase; specified material to be placed below the foundation of a structure.

as-built plans   construction plans modified to show changes made during construction.

asbestos cement pipe   formerly used extensively for water supply mains and sewer lines.

ASCE   American Society of Civil Engineers, 345 East 47th Street, New York, NY 10017.

ASCM   American Congress of Surveying and Mapping.

asphaltic concrete   material used for paving streets, parking lots, and sidewalks; composed of bituminous petroleum products mixed with aggregate.

ASTM   American Society for Testing and Materials, 1916 Race Street, Philadelphia, PA 19103.

axis of rotation   point about which the cross slope of a street is rotated (see Fig. 7.21).

azimuth angle   in surveying, an angle measured clockwise from a specified meridian.

backflow-prevention device   check valve to prevent sewage from backing up into the plumbing facilities within a building if the downstream sewerage becomes blocked.

back up   copying computer files for safekeeping.

balance   term used to describe the situation when the volume of required earthwork excavation equals the volume of required earthwork fill.

balance line   line at the volume of earthwork on a mass diagram where the excavation and embankment material are equal (see Fig. 6.20).

base map   plan showing only that information basic to several aspects of the design (actually not a map). Copies of the base map are made and information needed for various aspects is added to separate copies, thereby saving repeated drafting of the same information.

bench   horizontal shelf formed in a cut or fill slope; the bench reduces the risk of landslides and provides space in which to collect drainage for safe removal (Fig. 6.2). Also used to prepare existing natural slope to accept fill (Fig. 6.5).

benchmark   vertical reference point.

bend   a conduit formed with a curve of specified deflection (Fig. 10.3).

berm   raised area or curb.

bird bath   small puddle caused by insufficient slope or poor workmanship.

BM   benchmark.

BMP   best management practices for preventing water pollution.

bond   legal document to ensure that specific work will be performed.

BSL   building setback line.

bulkhead   retaining wall.

bulking factor   number expressing the amount of increase in volume upon excavation of densely packed earth.

CAD   computer-aided design.

CADD   computer-aided design and drafting.

capital improvements   improvements in the form of tangible objects such as buildings and highways.

catch basin   storm water inlet, usually located in the curb and gutter of a road (Fig. 9.9*b*, *c*, and *e*).

catch point   place where a cut slope or a fill slope meets natural ground (Fig. 6.8).

centrifugal   moving away from a center.

cfs   cubic feet per second.

cfm   cubic feet per minute.

check valve   device which allows fluid to flow in only one direction.

chord   nontangent line segment which cuts across a circle but does not go through the center of it.

CIP   cast-iron pipe; used where high compressive strength is needed because of shallowness or where tensile strength is needed for pressure lines.

CIPP   cast-in-place pipe; used where very large concrete pipe is needed.

Class 3 toxic waste dump   disposal site constructed with plastic or clay lining to prevent toxic materials from contaminating the ground and groundwater.

CMP   corrugated metal pipe; usually used for culverts and storm water outfalls. When CMP is used, it should be accompanied by required gauge.

CO   clean out. Same as flushing inlet (FI, Fig. 8.2). Provides access from the surface to flush out sewer lines.

compaction   causing consolidation usually of soil, AB, or AC to create a firm foundation.

concentric   having the same center.

condominium   complex of residential or office units which can be owned by individuals. Each unit carries a proportionate undivided ownership in the underlying real property.

conduit   pipe or tube for conveying fluid, or a structure containing ducts.

contingency an unforeseen or unanticipated occurrence. In cost and fee estimates, the contingency usually ranges from 5 to 30 percent of the total construction cost or fee, to pay for unanticipated work.

contour line joining points of equal elevation (Fig. 6.7).

contour grading method of design using proposed finished contours (Fig. 6.16). May be used for determination of earthwork.

coordinate numbers representing distances north and east of a specified reference point.

corporation cock valve for controlling the flow of fluid or gas from a main to a customer (also called a *corporation stop*).

cos cosine; ratio (in a triangle with one angle of 90°) of the length of the side next to the angle divided by the length of the hypotenuse.

cp concrete pipe.

CPM critical path method; a technique for determining scheduling for efficient coordination of work.

CPU central processing unit; the part of the computer that performs the calculations and logic.

crest vertical curve continuously bending line that provides a smooth change from a positive to a negative slope, from a positive to a flatter positive slope, or from a negative to a steeper negative slope (Fig. 7.10).

critical path the longest route through a critical path method diagram.

critical points locations or points of information that must be adhered to.

cross section a plane that cuts through at right angles to the horizontal; also refers to a drawing of this plane (Fig. 4.5).

crown highest point in the cross section of a street—usually at the center.

CSP corrugated steel pipe.

CTB cement-treated base.

cul-de-sac short street with a bulb for turning around at the end.

curb return curved section connecting curbs at intersecting streets.

daylight (1) term used in earthwork to designate the point at which a cut slope meets natural ground; (2) *to daylight*—to cut or fill to natural ground.

degrees designation for incremental division of the angles. There are 360 central degrees in a circle.

DEIR draft environmental impact report; preliminary document prepared for a state environmental protection agency describing the effects of constructing a particular project. The draft report is circulated among different agencies and made available to members of the public for evaluation and comment before the report becomes finalized.

DEIS draft environmental impact statement; same as a DEIR, except that the document is prepared for the federal Environmental Protection Agency.

departure distance in an east-west direction from one coordinate to another.

development plan   drawing showing lots with building footprints (outlines) and any additional information that will be useful, such as house-plan designations and exterior finishes; plans for finished grading; indications of setback distances for zoning; locations of driveways, water, gas, and sewer laterals, and mailboxes. Can be called a *site plan*. May be used to acquire building permits.

digitize   to translate two-dimensional information into an electronic format for use in a computer.

DIP   ductile iron pipe; used where extra compressive strength is needed because of shallowness, or where tensile strength is needed for a pressure system.

download   copy or move electronic information to another location.

drag   moving electronic information from one location in a file to another using a mouse.

drag and drop   same as *drag*.

drainage basin   area that contributes storm water to a particular waterway or point of concentration (Fig. 9.3).

drainage release point   horizontal and vertical location where drainage is released from one drainage basin into the next (Fig. 7.16).

dry well   a hole dug in the ground which is then filled with aggregate to provide small-capacity underground storage of drainage until water can percolate into the ground.

DTM   digital terrain modeling; a technique for translating three-dimensional topographic information onto a two-dimensional surface.

dust palliative   a method or material used to lessen or eliminate dust from becoming airborne.

easement   legal right held by one owner for a specific use of the land of another. Also the location or description of that right.

ED   extended detention; holding back of water for a period of time to allow suspended solids to settle out.

effluent   liquid flowing from some source.

EIR   environmental impact report; document prepared for a state or local agency describing the anticipated effects of construction of a proposed project.

EIS   Environmental Impact Statement; document prepared for the federal government describing the anticipated effects of constructing a proposed project.

E.I.T.   engineer in training; term given to the first half of the professional license examination for engineers. Also, a person who has passed the E.I.T. examination.

electroliers   street lights.

embankment   earth filled in and compacted to provide a stable, usable surface; or the material required for the embankment work.

EPA   Environmental Protection Agency; department of the federal government with the responsibility for safeguarding the environment.

ER   end of return; the BC or EC of a curb at connecting streets (Fig. 7.17).

et ux   and wife; Latin term used in legal documents, including land descriptions.

et vir   and husband; Latin term used in legal documents, including land descriptions.

event storm   a storm of an intensity so great that it occurs only once every 100 years is sometimes called a 100-year event. This terminology is used for various return periods.

EWL   equivalent wheel loads; term to describe characteristics of traffic.

excavation   cutting away and removal of earth to provide a usable surface; or the material resulting from the work of excavation.

fault   fracture of the earth's crust where slippage has occurred either horizontally or vertically.

fault zone   area on either side of an earthquake fault to be kept clear of structures because of the risk of earthquake damage.

fauna   the animals that inhabit a particular area and time.

fee   dollar amount for professional services.

FF   finished floor; usually accompanied by the elevation thereof. Found on architectural and grading plans.

FGI   flat-grate storm water inlet (Fig. 9.9a).

FI   (1) field inlet; same as flat-grate storm water inlet; (2) flushing inlet; a sewer connection to allow clean out of a line; same as *CO* (clean out, Fig. 8.2).

final map   subdivision, parcel, or Record of Survey map in its final form, recorded or ready to be recorded.

flora   plants that grow in a particular area and time.

flow line   path traced by liquid, usually along the invert of a pipeline, ditch, or channel.

flow rate   quantity of fluid passing a point during a specified time, for example, cubic feet per second (cfs), million gallons per day (million gpd or mgd).

footprint   outline of a building or other structure.

force flow   liquid moved through a conduit under pressure.

force main   utility conduit carrying flow under pressure rather than by the force of gravity.

format   the order and arrangement of a particular document.

fps   feet per second.

freeboard   distance between the water level and the top edge of the ditch or the top of the pipe where the structure will be at capacity.

french curve   flat drafting template of scroll-like curves; used to draw curves of varying radii.

french drain   a subterranean drainage ditch filled with permeable material to capture subsurface water. A perforated pipe is installed for removing the water.

GIS geographic information system; a map including a wide variety of information about a particular geographic area. Usually on electronic medium.

glitch any temporary or random malfunction or error.

gpd/cap gallons per day per capita.

GPS Global Positioning System; a method of surveying using satellites.

grad angular increment of a circle divided into 400 increments. Used with metric measurements in some countries.

grading ring circular device for bringing the top of a manhole or other subsurface structure up to the desired elevation or level.

gravity flow flow of liquid drawn through a conduit or along a channel by the force of gravity.

grid paper drafting paper on which a grid is printed to facilitate plotting of scale drawings.

grubbing clearing the ground of roots and stumps.

guinea 2×2 in wooden stake used by surveyors as a vertical reference point.

guinea hopper construction worker who reads information from survey stakes and calls it out to the equipment operator.

Gunite mixture of cement, sand, and water sprayed over reinforcing steel as a lightweight paved surface.

GV gate valve.

ha hectare; a metric unit for measurement of area of land. One hectare = $10\ 000\ \text{m}^2 = 2.47$ acres.

hard copy drawing on paper as opposed to on an electronic medium.

hard disk permanent magnetic disk within a computer that holds the permanent information stored.

hardware the computer itself and its peripherals.

Hardy-Cross solution a method using systematic, successive corrections for assumed flows in a piping system. Used to design water supply networks.

head loss loss of pressure, often due to friction or turbulence caused by bends and changes in a piping or channel system.

hectare metric unit for measurement of area of land, abbreviated *ha;* 1 ha = 100 ares = $10\ 000\ \text{m}^2 = 2.47$ acres.

HGL hydraulic grade line (Fig. 9.19).

HI height of instrument; surveying term that indicates the elevation at the line of sight on a transit, level, or other surveying instrument.

hinge point location where the slope of a cross section changes, usually at the start of a cut or fill slope (Fig. 6.8).

HP hinge point.

hydraulic grade line the level to which the water surface would rise if the system were open; a line connecting the water surface at each end of a system (Fig. 9.19).

hydraulic radius (R_H)

$$R_H = \frac{a}{p} = \frac{\text{cross-sectional area}}{\text{wetted perimeter}}$$

hydraulics   science of the mechanics of fluids at rest and in motion.

hydrograph   drawing of the water level, rate of flow, or rainfall intensity plotted against time.

hydrology   science of the natural occurrence, distribution, and circulation of water on the earth and in the atmosphere.

hypotenuse   longest side of a right (90°) triangle; the side opposite the 90° angle.

IDF chart   graph of intensity, duration, and frequency of rainstorms; used to estimate expected storm water runoff (Fig. 9.1).

impermeable   not permitting passage of fluid.

infiltration rate   time it takes for water to seep into the soil, or rate at which groundwater seeps into sewer lines.

Internet   the largest computer network in the world. It links millions of computer users.

interpolate   process for estimating an intermediate term; process for estimating an intermediate value by including a proportion of the difference between two values.

interstices   small empty spaces between parts, such as the spaces between the rocks in a layer of aggregate.

inundate   flood.

inversing   determining bearing and distance between two known coordinates.

invert   inside bottom of a pipe; the flow line.

inverted siphon   conduit curved concave down to circumvent an obstacle.

ionosphere   the outer part of the earth's atmosphere.

iterate   to do repeatedly.

jurisdiction   power or authority over particular areas, for example, the city, county, a flood control district, or a sanitation district.

key   keylock; groove or berm formed in material (concrete, earth, wood) to be fitted with a corresponding berm or groove on the matching layer to prevent slippage.

lateral   pipe connecting a utility main to facilities at the sides.

latitude   distance north and south between two coordinates.

lien   legal right to interest in property for payment of debts.

line of sight   a clear area where the eye can view with nothing in the way.

liquefaction   process of liquefying; term used to describe the condition some soils take on during an earthquake.

looping technique of connecting conduits so that few or no dead-end branches of piping exist.

L.S. licensed surveyor.

L.S.I.T. licensed surveyor in training; a person who has passed the L.S.I.T. examination, which is the first half of the examination for a surveyor's license.

LTB lime-treated base; material used to strengthen foundation material for structures.

magnetic north the north direction as determined by a compass rather than by astronomy. In deeds, north refers to astronomic north unless stated otherwise.

main primary trunk of a piping system.

mainframe large computer that can process information from several sources.

Manning's equation

$$V = \frac{R_H{}^{2/3}S^{1/2}}{n}$$

where V = velocity, m/s (or fps)
       n = coefficient of friction
       R_H = wetted perimeter, m (or ft)
       S = slope, m/100 m (or ft/100 ft)

map (1) plan-view illustration, usually drawn to scale, representing the relative locations of property, streets, or topography; (2) legal document to establish property exactly.

mass diagram chart showing the location of excavation and embankment activities and quantities (Fig. 6.20).

mass haul diagram mass diagram.

microfiche a sheet of film that can hold many documents which have been photographically reduced for viewing through a projector.

minute angular measurement: $\frac{1}{60}$ of a degree.

mitigation activities or structures to lessen the impact of some thing or event.

mobilization assembling equipment and personnel into readiness for some activity.

modem a device which interprets computers' electronic impulses into frequencies within the audio range, and vice versa, for the use of telephone lines to communicate computer information.

monument some tangible object used to identify a described location in surveying.

moratorium an authorized delay of some activity; for example, city governments may delay any further land development activities until more sewage treatment capacity is provided.

mouse handheld electronic device for use as a pointing and drawing device.

Mylar   filmlike material for drafting or printing reproducible copies.

NCPI   National Clay Pipe Institute.

NPDES   National Pollution Discharge Elimination System.

obliterate   completely destroy and clean something away so there is no evidence of it having existed.

off-site   areas or facilities outside the property lines or boundaries of a project.

on-site   areas or facilities within the property lines or boundaries of a project.

order of magnitude   relative size of a number.

outfall   place where a sanitary sewer line or storm drain line discharges.

overfill   (1) placing material beyond the required area, horizontally, usually to facilitate the use of equipment (Fig. 6.6); (2) placing material higher or deeper than required in order to promote compaction.

overt   inside top of a pipe.

PAC   political action committee.

palliative   material or method to minimize or eliminate an adverse effect.

parabolic curve   (1) curve used to form smooth transitions for vertical changes of directions for streets and pipes; (2) curve formed by the intersection of a right circular cone with a plane parallel with the side.

parcel map   map prepared to divide property into a limited number of units.

patio home   private residence which is placed along one property side line so that there is a usable side yard on the opposite side. The side of the house placed on the property line has no windows or doors.

PC   personal computer.

PCC   Portland cement concrete; concrete made with Portland cement as designated by ASTM.

P.E.   professional engineer; a person who has passed a licensing examination for engineers.

peak flow   maximum instantaneous flow.

peaking factor   factor to be applied to an average flow to yield a maximum instantaneous or peak flow.

perched water table   shallow water table above another larger water table (Fig. 6.4).

percolate   to pass through a porous material.

percolation pond   small body of water held to facilitate seepage into the groundwater reservoir.

perforated pipe   conduit with holes to allow groundwater to seep into the pipe so that it may be transported to a drainage system.

permeable   capable of being penetrated through pores.

PERT   Program Evaluation Research Task; a method for scheduling and evaluating construction progress.

PI   point of intersection.

piezometric slope   slope of water surface necessary for it to flow.

plan   design or list of steps to accomplish some task; usually, drawings made to scale that show necessary information for construction of structures.

planimeter   instrument for measuring plane areas.

planned development   project put together to provide amenities in a creative way.

plasticity   degree to which material is capable of being molded or shaped.

plume   underground area containing adulterated or toxic water fanning out from a source.

PLS   professional land surveyor.

PMP   perforated metal pipe; used to collect and drain underground water.

PMS   plant-mixed surfacing; asphaltic material mixed at a central plant to be placed as roadway or for parking surfaces.

POC   point on curve.

point of concentration   location where drainage comes together, such as a storm water inlet.

POT   point on tangent.

potable   safe to drink.

precision   degree of exactness.

PRC   point of reverse curvature.

precipitation   water falling to earth in the form of rain, snow, sleet, or hail.

profile   line representing a longitudinal, vertical section through a roadway or pipeline (Fig. 4.5).

profile paper   drafting paper with a grid printed on it to facilitate plotting.

PS   protective slope; earth covering a foundation. Usually shown with an elevation at its highest point for comparison with surrounding elevations.

PS&E   plans, specifications, and estimates; final construction documents.

public works   governmental facilities.

PVC   polyvinyl chloride pipe.

Pythagorean theorem

$$h^2 = a^2 + b^2$$

where h = hypotenuse or longest side of a right triangle (90°)
  a = one leg adjacent to the right angle of a right triangle
  b = other leg adjacent to the right angle of a right triangle

Q quantity of flow; usually in cubic meters per second (cubic feet per second, cfs) or million gallons per day (million gpd or mgd).

quitclaim deed   legal document whereby makers release any hold they may have on a particular property. Often prepared where the maker's claim is not clear but of sufficient significance to cloud the title.

ramp   sloped passage way for wheelchairs and pedestrians. Traveled way for vehicles to enter freeway.

rational formula

$$Q = CIA$$

where Q = quantity of runoff, m³/s (cfs)
$\quad\quad C$ = runoff coefficient
$\quad\quad I$ = rainfall intensity, mm/h (in/h)
$\quad\quad A$ = area, hectares (acres)

RCP   reinforced-concrete pipe.

R.C.E.   registered civil engineer; a person who has passed the licensing examination for civil engineers.

R.E.   (1) registered engineer; a person who has passed the licensing examination for engineers; (2) resident engineer; a person stationed at the job site to ensure that the work is performed according to the plans and specifications.

release point   horizontal and vertical location where drainage is released from one drainage basin into the next (Fig. 7.16).

retention pond   basin to hold storm water runoff and to provide a gradual release.

return period   amount of time between storms of a particular intensity.

R_H hydraulic radius; distance in a channel or conduit that will be wet when flowing at a particular rate.

right-of-way   strip of property provided for a street, highway, or other linear facility.

rubberbanding   computer graphics term for when one end of an object stays in place and the rest of the object is stretched to another location.

runoff coefficient (C)   number representing the amount of water running off an area as a proportion of the amount falling on it, based on type of soil, surface coverage, and evenness and degree of slope.

sag vertical curve   continuously bending line that provides a smooth change from a negative to a positive slope, from a negative to a flatter negative slope, or from a positive to a steeper positive slope (Fig. 7.10).

scale   (1) measuring stick. A metric scale is divided into 1:20, 1:25, 1:50, 1:75, 1:100, 1:125. On an engineer's scale, 1 in is divided into increments of 10, 20, 30, 40, 50, 60, or 80. On an architect's scale, 1 in is divided into ½, ¼, ⅛, 1/16, 1/32, 3/32; (2) to measure using a scale.

scanning   a technology that allows visual information to be transformed into electronic information for use in computers.

screening   (1) to cause material to pass through screens; a method for segregating sand and aggregate by size; (2) photographic procedure to produce faded lines.

second angular increment of a circle; there are 60 seconds (60″) in an angular minute.

septic tank underground holding tank to allow organic sewage to separate and decompose. It is connected to leach lines that draw off the purified liquid for percolation into the soil. Used for residences or other buildings where there is little sewage and no public sanitation facilities.

setback distance from which structures must be kept clear.

sewage waterborne waste carried in sewers.

sewerage network of sewer lines.

sheet flow flow of liquid moving evenly over an area without being concentrated in swales.

shrinkage decrease in volume caused by compaction of soil.

significant figures all the nonzero digits of a number and zeros at the end that show degree of precision (also called *significant digits*).

sine (sin) ratio (in a triangle with one angle of 90°) of the length of the side opposite the angle divided by the length of hypotenuse.

slope degree of rise or fall of a line with reference to the vertical and horizontal planes (Fig. 6.1).

software instruction information for the computer. There are two types: (1) system software, made up of operating systems, communications software, and database managers; and (2) applications software, the specific programs the user chooses to accomplish particular tasks.

specifications written instructions of methods and materials to be used to produce a product.

spreadsheet a matrix of rows and columns used for a wide variety of calculations and comparisons of information.

standpipe vertically mounted pipe open at the top, with or without a terminal elbow (Fig. 3.6).

station designation on a centerline or reference line at 100 m (or 100-ft) intervals.

station yards a one-yard volume of earth carried one station (100 ft).

stereoscopic viewer optical device for viewing, in three dimensions, pairs of special aerial photographs taken for this purpose.

storm water release point see **release point.**

stratosphere atmospheric zone reaching from the troposphere to the ionosphere.

subdivision parcel of land divided into lots for legal descriptions and real estate transactions.

subgrade material placed below the surface to provide a stable foundation for structures.

subsidence settling of the earth's surface due to excessive removal of underground material such as groundwater.

superfund site   area identified by legislation as being sufficiently polluted to be a threat to the health and welfare of people and eligible for federal funding to accomplish cleanup.

superelevation   tilting of the cross slope to compensate for centrifugal force or to remove drainage (Fig. 7.21).

swale   small valley area between two hills or mounds which is lower at one end. The swale provides a pathway for drainage.

SWE   Society of Women Engineers, United Engineering Center, 345 East 47th Street, New York, NY 10017.

SWI   storm water inlet; includes catch basins and field inlets (Fig. 9.9).

SWPPP   Storm Water Prevention Pollution Plan.

systems engineer   a person who matches and coordinates the computer hardware and software to satisfy particular requirements.

tangent   (1) straight line or flat plane that touches a curve at a single point; the line or plane will be perpendicular to a radial line at that point; (2) (*tan*); the ratio (in a triangle with one angle of 90°) of the length of the side opposite the angle to the length of the side adjacent to the angle.

TBM   temporary benchmark. See **benchmark.**

t_c time of concentration.

TC   top of the curb.

template   an accurate copy of some shape that can be used repeatedly.

tensile   capacity for stretching.

TI   traffic index; numerical value to indicate expected usage for design of pavements.

title report   description of ownership of property, easements, and liens against it.

topo   topography.

topography   physical features and relief of a site; includes buildings, trees, utilities, paved areas, waterways, and elevations.

townhouse   single-family residential house which is attached to adjacent houses. The building includes ownership of underlying property.

township   an area of land approximately 6 mi square and designated by its relative distance north or south from a baseline and its relative distance east or west of a meridian. The relative distance from the meridian line is described by number of ranges (6 mi) east or west of the meridian line.

toxic plume   underground water containing toxic substances; usually occurs in the shape of a plume fanning out from the source.

tract map   subdivision map given a tract number to facilitate referencing.

traverse   a series of courses of lines and curves, usually described with distances and bearings.

trench spoil   earth dug out of a trench.

triangulation   technique for determining distances between points and verifying locations of points.

tributary   making a contribution.

troposphere   the earth's atmosphere from the surface to the stratosphere, reaching 9.6 to 19.3 km (6 to 12 mi).

true north   astronomical north, as opposed to magnetic north.

truck turns   graphic representations drawn to scale of the tracks of front and rear wheels for various size trucks and other large vehicles.

truncate   to cut off; a number is truncated after the necessary number of significant figures.

TS   top of slab; usually accompanied by the elevation thereof. Found on architectural, grading, and development plans.

TTN   topographic triangulation network; used as a graphic method to illustrate three-dimensional information.

twist   rotation of data while maintaining coordinates in Autocad.

UBC   Uniform Building Code; manual of specifications for building design.

undercut   excavating less in a horizontal direction than is called for. Usually done at property line or next to building pad (Fig. 6.6).

unit cost   cost associated with a single item.

UPC   Uniform Plumbing Code; manual used for plumbing specifications.

U.S.A.   Underground Service Alert; agency that will have underground utilities located and marked on the surface to prevent damage during construction activities.

USC&GS   U.S. Coast and Geodetic Survey.

USGS   United States Geological Survey.

valley gutter   concrete swale down the center of a roadway or across a secondary street (Figs. 7.24 and 9.10).

vara   historical method for measuring land. 1 vara = 33 inches.

VCP   vitrified clay pipe; usually used for sanitary sewers less than 1070 mm (42 in) in diameter.

vellum   high-quality, translucent drafting paper.

vernier   a short graduated scale which slides along a longer scale, devised to facilitate measuring.

virus   electronic bug that corrupts electronic files.

vitrified   treated with heat to cause to be glasslike.

w/   with.

web page   information presented in text and graphic form on the Internet.

weighted average   average value based on relative significance of the items being averaged.

wet well   chamber that fills to a designated level with fluid, at which level a pump is activated.

wetland   area of land covered all or part of the year with water and providing an environment for certain species of flora and fauna.

wet tap   connection to a pipeline which is carrying fluid.

wetted perimeter   length of surface measured on the cross section that will be wet when ditch or pipe is flowing at a designated depth.

windrow   earth placed in a linear pile.

wye   Y-shaped pipe connector.

Index

ABOUT THE AUTHOR

Barbara Colley, P.E., is the head of BC$_3$E, Barbara Colley Consulting Engineers, a firm that provides a full range of civil engineering design services, based in San Jose, California. She is a land development specialist with over 30 years experience.